Lulu Press

The Rape and Murder of Mother Nature

By the same author:

The Gardeners of Eden	George Allen and Unwin, 1973
Eyelids of Morning	with Peter Beard. The New York Graphic Society, 1973
Suleiman's Gold	The Book Guild Ltd, 2010

The Rape and Murder of Mother Nature

Alistair Graham

Lulu

First published in the United States in 2022

ISBN 978-1-716-14817-0

Keywords

abstraction, afterlife, aggression, belief, brain enlargement, burial, civilization, coercion, conscience, cooperation, copulation, cultural modification, defense mechanisms, economic expansion, ego, fear of death, group behavior, guilt, human evolution, human instincts, immortality, inclusive fitness , leadership, magic, natural selection, nature, ownership, Pleistocene, pleasure principle, population, religion, rhythm, slavery, supernatural

Printed in the United States by Lulu

This book is dedicated to the memory of Mother Nature

Looking back down the foggy ruins of our time
All those millions and millions of long years ago
Dear Mother Nature did guide us, keep us in line
With evolution's dictates of natural selection
That shaped us, molded us and made us to shine,
While others in Eden danced their Last Tangos.
Selection favored a super- complicated mind
That puffed up our egos and dreams of more, more.
Noah's Great Rainbow convinced us we were sublime
The Pinnacle of all Life, radiant Divine Confections,
Masters of the Universe and Ringers of the Chimes.
We forgot who we really were, we thought
We were big deal, omnipotent, capable of anything.

Then we turned on Mother Nature we treacherous swine,
Saying it's Manifest Destiny that we sweep you away.
We don't need you, neither you nor your design.
We've so much better plans for all that wilderness
Than tolerate your vermin that harass and combine
To hinder our vision of a world that's far above you,
That our wisdom assures us will eventually define
Our infinite happiness and wealth beyond measure.
The Idiot Wind that's howling we take for a sign
That what we do's amazing, so clever and inspired.
We deny all our treachery, our monumental crime,
For we are ignorant and arrogant, stupid and blind,
Incapable of thinking beyond our selfish little minds.

Preface

Those who study the human mind might rightly ask on what ground do I think I can explain how it works. There is a vast body of scientific (and unscientific) fact, hypothesis and speculation on the mind, already too large by far for most individuals to be familiar with all of it. So why me? Good question. Moreover, it can reasonably be said that as much as we know it is paltry compared to what we have yet to know. There is a long list of mental phenomena whose mechanisms are unknown even if we can see them operating easily enough. An odd paradox – to be overwhelmed simultaneously by too much and too little knowledge.

At the same time as I accept the daunting amount of information out there, and the intrinsic complexity of the material, I believe it is pointless if one cannot synthesize it in such a way as to be intelligible to the average educated, thoughtful person. I'm confident that what we know is enough to grasp the salient features of our behavior and understand its seemingly bizarre perversity. There is also the glaringly obvious fact that our behavior as a society in no way reflects either our accumulated wisdom or our potential sagacity. On the contrary, it is astoundingly simple, crude and short-sighted. We devote our all to the pursuit of pleasure, gift-wrapped in spurious political and economic rationalizations, tarted up with virtual ideals of peace, prosperity and freedom, of equality and justice, that dance before our gullible minds like the phantoms of mirage – hazy illusions, always there, never attained. The point being that understanding our main behavior does not call for any inaccessibly deep or esoteric enquiry. The basic workings of our minds will suffice. That said, even the basics call for a fair bit of information. It's not, as the Nigerians say, a one-day job.

For the most part I've not referenced this work, partly because it's not a scientific treatise, even though it's as accurate and up-to-date as I can make it. There is also the problem that as an independent scholar not associated with an institution I would have had to pay to read many references, making them inaccessible. Properly crediting the originator of much material requires vast amounts of bibliographic research – facilities and time I don't have. Even though the bulk of assertions I make are unreferenced they are all drawn from reputable, well-known sources. I have tried to make it obvious whenever an assertion or hypothesis is mine and not supported by research. The general reader who wants to know more about some aspect of our behavior need go no further than the well-known, rapidly improving encyclopedias such as

Wikipedia, Encyclopedia Britannica and others. Look up almost any topic and there will be excellent books on it.

Outside the technical literature one has only to read Tom Wolfe's brilliant *The Bonfire of the Vanities* to find what I am trying to say about our ego-driven, impulsive behavior already eloquently said without any need to consult science. Irvine Welsh's *Glue* and Erica Jong's *Fear of Flying* fill any gaps left by Wolfe. Not surprisingly, creative writers have always seen what we do for what it really is, because creativity springs from the deepest parts of our mind. The Old Testament prophets and their Mesopotamian predecessors also lyrically spelled out much of our egoistic behavior and society's idealistic attempts to manage it, albeit with a god-fearing bias.

The more formal explanation of what makes us do what we do has had, more than most aspects of zoology, an unfortunate history, mostly owing I think to the realization of everyone who has ever tried to understand the mind that there are seemingly inscrutable causes of so many straightforward effects. So much of what we think and do is at odds with what we would like or expect or hope for. What we try for is so often balked by our mental waywardness. We easily think of solutions to problems, but lack the will-power to implement them. Apparent explanations usually turn out superficial. The unfortunate part of it is that the study of the mind – psychology - was not begun scientifically, but philosophically. And philosophers generally are not qualified for such studies because theirs is an inward-looking world unconcerned with the outward realities of Nature. As Cicero remarked more than two thousand years ago, '*Nothing so absurd can be said, that some philosopher has not said it.*' I can assure Cicero that they keep trying.

Psychology just can't seem to shake off the leaden cloak of transcendental moonshine that smothers plain, simple science. One has only to read the Oxford Companion to the Mind (which also contains many excellent essays) to see how much space is still wasted on philosophical maundering in what is meant to be an authoritative technical reference book. No other branch of science does this. Physics and zoology for example have enduring Standard Theories against which one can hold up pretty well everything one investigates without foundering on intractable contradictions. In a nutshell, that's what's wrong with psychology; the failure to match psychic observations against zoology's standard theory of Darwinian evolution. The theory has lost none of its original power to model animal characteristics, mental as much as physical, in terms of their environmental adaptiveness. There is no reason not to accept that everything about our minds has evolved in accordance with Darwinian evolution. The detailed mechanisms may be – often are - difficult to unravel, but that is no reason to look for other explanations.

****

In the process of trying to figure out all these things I often found myself going back over experiences that, trivial enough in themselves, nevertheless contributed to my understanding of human behavior in many important ways. These recollected experiences always left me feeling that I had been lucky to have had them ....

Lucky that I spent so much of my life among wild animals, that I have lived in parts of the world such as the east shore of Lake Turkana that were at the time still free of modern, urban man and his crass manner of deconstructing Nature.

Lucky that I have lived among people who went naked, or all but, all the time, and to find that they were the same as us – just less affected.

Lucky that I hunted wild animals and knew what it feels like to be attacked by an animal bent on my violent death. And to know that my emotion on such occasions must have been much weaker than had I been armed with only a stone-tipped wooden spear.

Lucky that I spent enough time with such animals as elephants and chimpanzees to know that we are most definitely not the only sentient animals, and certainly not the most attractive.

Lucky that I was born and grew up in Kenya and was thus able to catch a glimpse, however fleeting, of the apes and other animals that inhabited the last vestiges of The Garden of Eden, the Pleistocene Paradise that my fellow ape-men were busy ripping apart.

Lucky that as a child I watched two old, naked Nderobo men who had heard that a rhino had just been killed rush up to the carcass, jump on top knives in hand and frenziedly cut pieces off and eat them. I particularly remember the pure bliss that emanated from them; their absolute ecstasy.

Lucky that when I left the Botswana Parks & Wildlife Department in 1979 the director, Kukes Ngwamatsoko, wrote me a nice farewell letter. 'Take care,' he said on hearing that I was going to Papua New Guinea, 'they eat people over there.' He wasn't joking – he really did think I should watch my back.

Lucky that I stopped once at a roadside café in a village near the Cross River National Park in Nigeria. My driver, Patrick Elephant, remarked that were he not in the company of whites (my wife and I) he most certainly would not have stopped in this village. Why not? Because there are cannibals here always on the lookout for strangers. If they think nobody will miss you they will kill and eat you and make medicine from your bits. They store extra meat by encasing chunks in balls of mud that are baked to make confections that have a good shelf life. But, said Patrick Elephant, he was safe in our company, because the cannibals only go for really black, black people; they scorn the insipid flesh of whites.

Lucky that I saw that *Jua Kali* (literally Hot Sun) village artisans in East Africa treat cuts and scratches with battery acid, equating the pain of the treatment with the intensity of its curative powers.

Lucky that I lived through the Mau Mau episode in Kenya and saw the macabre ease with which a normal human can be turned into a remorseless killer of his own kin by nothing more than a ceremony; and then exorcised into the original normal person by the same, but antithetical ceremony.

Lucky that I saw how many Africans admired the bloodthirsty slaughters of his own people by Idi Amin in Uganda, not because they endorsed the killings themselves, but because he wielded such enormous power over his people, by, it seemed the sheer force of his braggadocio. "He's such a bull" they'd say "Such a big man."

Lucky that my parents were wise and generous enough to persuade me to go to a university where I had the benefit of an education in classical evolutionary zoology.

Lucky that I was and always would be a Gardener of Eden, unashamedly opposed to the rape and attempted murder of old Mother Nature for no more reason than that I cannot see mankind will gain what he thinks he will from such a deed. To believe, as most people seem to, that we have reached a point where we can do without Nature, that we are above such barbarity, is a sad delusion.

Lucky to have been born the year before Freud died; I was alive in his lifetime. Meaningless, of course, but pleasing to me.

The following abbreviations are used:
Ya – years ago
Kya – thousands of years ago
Mya – millions of years ago

CONTENTS

Lulu Press i

Dedication v

Preface vi

The Rape and Murder of Mother Nature 1

Introduction..1

Part 1. How the mind works...................................5

Chapter 1. The Solution to Why, Oh Why 5

1.1 The unconscious mind 9

1.2 The Conscious Mind 12

1.3 Antithetical nature of things 16

1.4 The Pleasure Principle 16

1.5 Defense Mechanisms.....................................19

1.6 Idealization..25

Other Features of the Freudian Model 28

1.7 Aggression...28

1.8 Copulation 32

Chapter 2 Group Psychology.......................................36

Chapter 3 Neurons and the Brain 41

The Brain 41

The Neuron 45

Chapter 4 Genes and Natural Selection.....................51

Part 2 Our early evolution 59

Chapter 5 The Miocene and Pliocene 59

The Miocene 60

The Pliocene 62

Part 3 65

Chapter 6 The Pleistocene 65

Part 4 79

Chapter 7   The Big Beautiful Brain 79

Part 5 87

Chapter 8   Belief and the Supernatural 87

    8.1   Religion 96

    8.2   Rhythm and music 104

    8.3   Our Conscience 105

Part 6   The last days of the Pleistocene 108

Chapter 9 The Rape of Mother Nature.................................. 108

Part 7   The Holocene 121

Chapter 10   The Murder 121

    10.1   Ownership, Leadership and Slavery 130

    10.2   Unification, Coercion and Cooperation 132

    The Myth of Cooperation 132

    10.3   Reading, writing and arithmetic.............................136

Chapter 11   After all, we are only animals 139

    Acknowledgements.. 1

    About the author 2

The Rape and Murder of Mother Nature

Reason – the power of the mind to think, understand and form judgments logically

Introduction

Why, Oh why? For much of my life I've asked myself a thousand times over why, why, why? Why do we do such bizarre and contradictory things? Most are trivial enough, but many are the most fundamental of things, the things that determine the destiny of all mankind. Considering what we know about our behavior, and given that if we make the effort we can analyze things objectively, I was bewildered by the endless repetition of the same foolish behavior that never changed. Over the last three centuries our scientific knowledge of ourselves has exploded, but our use of it doesn't show when it comes to how we decide what action is going to be in our best interests In fact, the opposite is a far more rational conclusion to draw, as has been pointed out by many people many times. We use our brains to wreck our opportunities far more than to run with them. Creativity is stifled by crassness.

We build vast agricultural industries wholly dependent on rain-fed farming, and react to each inevitable drought (or flood, or fire) as if we had been taken unawares by some malevolent force, dealt an undeservedly wicked blow – victimized. We build homes and even towns on active faults, on floodplains, at the feet of volcanoes, on the sea shore, downstream of dams, in highly flammable forests and tornado and hurricane zones. We know – because insurance actuaries accurately calculate the probabilities – that catastrophic eruptions, sea level rises, earthquakes, famines, fires, floods and winds will cause havoc, chaos, panic and pain to the inhabitants from time to time. Each inevitable time, there's much weeping and wailing and wringing of hands, and commissions of enquiry that make long lists of recommendations to avoid everything – except the real causes. The treachery of Nature is invoked, the inaction and indifference of government is denounced, money and sympathy are beseeched by the 'victims' and after the event anniversaries are somberly observed 'Lest we Forget', the irony unnoticed. For the slogan should surely be

'Lest we remember', because if we did remember what really caused the event, and used the memory to avoid making the same mistake again ....

But, that's not how we use these swanky brains of ours. We do nothing until the inevitable repeats itself, regardless of how obvious the inevitable was. Then those who avoided the risks are made to pay for those who didn't. And then we stubbornly rebuild and re-inhabit the same old places and wait for it to happen all over again. And again. And again. There's no sense in it, it's plain lunacy, yet we can't stop ourselves. The only way such contrariness can be understood is to assume that some powerful mental process overrides common sense, in spite of the latter's acknowledged utility. Just what that process is, why it evolved and under what circumstances it's activated is the purpose of this enquiry.

The above are trivial examples compared to the really big why-why-whys, the monsters that are shaping our destiny. I mention the little things because they are a good way of introducing the far bigger, more dangerous and less obvious things. Anybody can see for themselves the paradox of, say, imposing speed limits while encouraging the manufacture of flimsy automobiles that can be and are driven at twice or three times the permitted speeds by people incapable of controlling and vehicles at such speeds – even when sober. As with the big things, we don't let such contradictions bother us. Most people shrug their shoulders and dismiss such thoughts. What's on television is as much as most people want to grapple. Which is okay as long as those whose job it is to deal with such matters – our leaders – pay attention and act. But our leaders are primarily concerned with holding on to power rather than making difficult, hard decisions.

So what are these monsters whose fiery breath like the climate is getting hotter and hotter? The latter-day dragons that even Saint George would wet himself before are quite simply our furious, headlong pursuit of ever-expanding economies, supercharged by relentlessly expanding populations. Why is everyone so hopelessly bogged in this extraordinary behavior? Why are our leaders so sure of themselves that nothing will induce them to reflect upon such matters, to wonder why they take such things to be set in mental concrete? Why do they absolutely refuse to consider the glaringly obvious consequences, to ignore reasoned dissenting voices, to act as if there simply is no alternative; that what they are doing is not just the *right* thing but the *only* thing?

It's not as if no one has any qualms. Many do, and have shouted them as loudly as they can for as long as the phenomenon has been recognized. But, the individual in our so-called democratic societies is an outsider if he questions society's treasured shibboleths. In fact, the louder he shouts the more he inadvertently hardens the very belief he denounces. Try sowing the seeds of doubt in the collective mind and you reap intransigence, not sympathy. You cannot destroy absolute conviction without offering a more attractive, more seductive alternative. Above all, you cannot appeal to reason where belief prevails. Belief in fact is our answer to reason where the latter fails to deliver a satisfactory explanation. By satisfactory I mean to the individual in question, not necessarily to everyone, or to society as a whole. The only way an objectively reasonable person can reconcile belief with reason is to work out why the propensity for belief evolved into one of our most robust mental functions, one that has profoundly influenced the course of civilization.

Finally, I realized that the only explanation for my bewilderment had to be that some vital element was missing from my own thinking. Somewhere along the line, I was assuming too much (or too little) about how people think. In the end, it wasn't difficult to pin down the dark matter that was fouling up the equation, or to see why it's so dark, so hard to see.

My purpose in writing this is to unravel the mental basis of why we have this all-pervading hunger for more, more, more, and to show that it's on the one hand simple enough to see why, and on the other a complicated product of our evolution from an unassuming chimp-like ape to a voracious, pig-headed and dissatisfied one.

There are two distinctly different aspects of human behavior to be considered when trying to work out why we behave as we do: individual behavior; and group behavior. Once individuals form groups their behavior changes profoundly and while the two manifestations spring from the same source the differences are extremely important and must be clearly understood before there's any possibility of understanding society as a whole. An understanding of either one however deep, without understanding the other has no chance of explaining the whole.

As groups are composed of individuals it's necessary to begin by explaining individual behavior, even though somewhat paradoxically we cannot understand the human condition as a whole until we see how we behave in groups. I hope to make this clearer as we go on.

I've divided this book into seven parts. Part 1 deals with the essential components of our behavior, plus a brief description of our brain and the way in which our genes under the influence of Darwinian natural selection determine what we are. The remaining parts trace our evolution as a species over the 19 million years since our ancestors, the first apes, evolved. Part 2 summarizes the little we know of the first 16 million years. Parts 3, 4 and 5 cover the Pleistocene (to 600,000 years ago), the period during which our evolution came up with much of what makes us today. Part 6 covers the last 600,000 years of the Pleistocene and the 35,000 years that brought us to the Holocene. Part 7 deals with Holocene – the contemporary era.

Part 1 . How the mind works

Chapter 1.   The Solution to Why, Oh Why

To us he is no more a person
Now but a whole climate of opinion
Under whom we conduct our differing lives.

Auden in memory of Freud

In trying to work out what the dark matter was that jammed my attempts to understand why we seemed unable to use the considerable capacity of our brains, I chose to use Freud's psychoanalytic model[1] of how our minds work. This needs a word of explanation, because it's fashionable among those who study our behavior to disparage the Freudian model as too speculative, or outdated, or unscientific. If they had replaced it with a better model, there would be no discussion; but they haven't. Some modifications have been advanced; many of Freud's interpretations are better discarded. Yet, in spite of the vast amount of research since Freud's day, his key observations and the insights he came by as a result remain as valid now as when he advanced them.

It was Darwin who saw that all animals including us were animated by instincts, behavior that arises spontaneously and inevitably because we are born with it. He realized that instinctive behavior was simply one of the many adaptations to the environment that an animal acquired as a result of natural selection. In this sense, it was no different to the physical adaptations that make up our bodies. Freud made his observations on human behavior with this Darwinian view in mind, which he took for granted as essential to any interpretation of why we behave as we do; namely that our instincts initiate everything we do, even though we have evolved elaborate mechanisms for modulating those instincts.

Freud's wariness of publicity, his refusal to act the celebrity was certain to put people off, to want to ignore him and his ideas. We – scientists and believers alike - demand heroes, because we want to see our own ideals enacted in those we admire. His interpretations painted a picture of us as

---

[1] Rather, the model Freud's insights would establish; he didn't present a model as such.

compulsively narcissistic, priapic, aggressive animals handcuffed to unconscious drives that blinded us to any hope of attaining the high-minded ideals we so dearly wished to believe we were pursuing. As if that was not insulting enough, he defined religion and belief in god as a primitive, infantile illusion that severely hindered much of what civilization was all about. Creative people already sensed these things and were not offended. Others were intimidated by such base allegations and preferred to dismiss or deny them. A Jewish psychology, as many said (and still do), by which of course they meant an obnoxious psychology.

Insofar as scientists are, or are supposed to be objective their dismissive attitude to the Freudian model is mainly owing to the fact that Freud's insights are wholly the result of observation, not experiment. That means they are not classically provable or falsifiable, and are therefore, for the conventional scientist, unscientific, and cannot form the basis of investigation. Formal research into how the brain works thus proceeds, rightly enough, without reference to the Freudian model. It takes what can be demonstrably shown and proceeds from such starting points alone. As what can be currently shown is only a tiny fraction of what is happening in our minds there's no way a comprehensive, truly scientific theory of human behavior can yet be constructed. We are light years away from that. Most research is still at the initial stage of the functional anatomy of neurons, and tortuous behavioral experiments constrained by the ethical barrier of direct intervention. Because the brain and nervous system are electro-chemical structures the physical anatomy gives little clue to how they work. That can only be investigated by immensely difficult observations and measurements of events at the organ, cellular and molecular level that turn out to be every bit as complicated as would be expected from our complex behavior. What this research, that in time to come will be seen as the early descriptive stages, does already show is that the functional anatomy of the brain is consistent with the sort of mechanisms that could be expected to underlie the generalizations of the Freudian model. There are several aspects that begin to provide a sound scientific confirmation of some of Freud's most important insights. More of this later.

For all that scientists must adhere to the scientific method's principles, they sometimes forget that the stringent demands of their profession leave them floundering in the face of such everyday human instincts as belief or guilt or sorrow; science simply can't yet define the neural mechanisms involved, and for the most part simply ignores such matters. For the time being, we have to go by observation knowing that our interpretation cannot be complete. But, we are dealing with behavior that everybody is familiar with – and has usually experienced. As long as observational interpretation sticks to what we do already know about the electro-chemical operations of the brain, and to the evolutionary genetics behind it all, we are likely to approximate the truth. It's when interpretation wanders off the beaten track of reality into such flimflam as mind-body duality, metaphysics, planes of being or purely philosophical trains of thought that we will go wrong. For sure, both psychoanalytic observation and scientific demonstration must one day arrive at the same destination (as Freud himself remarked) before we can be certain of the whole truth. But, that end-point is over the horizon when it comes to the mind and on the long voyage there we must make the best of what's on offer.

So, back to the Freudian model. The over-riding conclusion that Freud drew, and the reason why he devised what he called psychoanalysis, is that our overt behavior is a façade, so to speak, designed to present a sanitized picture of what we do and why we do it. Behind the façade lie motives that are often too primal or forbidden or uncomfortable to be directly expressed - as much for the presenter as for the receiver. Freud realized that these underlying motivations could only be teased out by painstaking analysis of overt behavior. He further realized that there are sound reasons why we must sanitize our basic urges, reasons that lie at the heart of everything we call civilized society. And finally, he saw that if we are to fully civilize ourselves we must before anything else understand what lies behind the façade. We must know why we behave as we do rather than take for granted that what we do is fully explained by what we claim to be doing. For a zoologist, this concept lies comfortably with the theory of evolution by natural selection, which has proved one of the most robust and useful general theories ever produced by science. For a zoologist there's nothing about our behavior, neither the outward façade nor the cryptic reasons behind it that cannot be explained by evolutionary adaptation.

> When did reason ever direct our desires or our fears?
> *Juvenal*

In what follows I will try to show why I have these convictions by tracing our evolution and the light it sheds on our behavior and why it evolved the way it did. No other animal modifies its innate behavior in the way we do. It's only the various instruments of modification that have made our behavior so difficult for so many to understand, and which have caused so much speculation about our place in Nature. We are certainly complicated organisms and to explain what I mean by all the above I see no way but to go through step by step the important elements of our makeup.

Freud could not see any fundamental difference between sexual pleasure and any other sensual pleasure, i.e., *all* sensual pleasure is simply graded from slight – scratching an itch- to strong – copulatory orgasm. As he also observed that the self engages exclusively in behavior that pleases it he could not but conclude that it's our sexuality that pushes us into so much of what we do. The logic is incontestable, whether or not one attributes the same importance to it as Freud did. He adduced as evidence the fact, which he drew from Darwin, that we are animals, not some unique organism. As such we are primarily motivated by instinct, the ancient genetically controlled behavior of our unconscious mind. Where we differ from other animals is in our cultural modulation of instinct. As spontaneous copulation, for example, is anti-social in our society the urge to it is held at bay and only (ideally of course) allowed out under culturally acceptable conditions.

Whatever reservations about Freud's model of our minds people may have the fact remains, as Auden so neatly put it, that his basic insights underpin contemporary Western society's understanding of itself and of the individual's behavior. The functions of the unconscious, the conscious and the conscience, the ego and its despotic self-importance, the impossibility of quelling instinct, the overwhelming pursuit of individual pleasure, the fragility of reason, the instinct for belief; all these and other psychic truisms have seeped into general awareness, albeit in dribs and drabs rather than in their entirety. While Freud's invention of psychoanalysis as a therapy does not always do better than other treatments, and is inaccessible to most, it has greatly eased the personal battles of countless people, specially the creative. More than that, it was by psychoanalysis that Freud gained his most penetrating insights into how we behave and why.

Freud began with the basic observation that the apparatus of our minds is stratified. The different roles played by the various elements cannot be understood as acting in isolation, even though we have to describe them one by one. Rather they act as 'layers' of mentally superimposed agencies so that there's a hierarchy of thought governed by stimulation and inhibition of stimulation. A sort of binary system where every source of mental output is connected to an inhibitory system that controls the intensity of the output – or even smothers it altogether.

With that simple model in mind he showed that regardless of the mind's obviously huge detail and complexity it can be understood well enough by dissecting out its main functional elements and showing how these act upon each other to attain the best outcome for the individual in plain evolutionary terms. He began life as a neurologist and knew better than most that the corresponding parts of the brain were unknown in his day and so avoided speculation on that aspect, just as Darwin and Wallace did when positing the agencies of natural selection. Today, the functional anatomy of the brain is beginning to emerge and the basics of Freud's model of the mind well match the way in which the nervous system has so far been seen to work.

Freud showed above all that the mind functions unconsciously as well as consciously; and that the conscious is for the most part unaware of what the unconscious is doing, even though the conscious is largely – perhaps entirely - motivated by the unconscious. This single fact is the cause of much of the difficulty we have had in fully understanding ourselves. It's a strange thing that we all have the same brain that allows us to assure one another that we all know how everyone behaves because we are all looking at the same behavior – in ourselves and others. At one level, it's true that we are all familiar with human behavior, that a lot of it is obvious, easily understood and universal. It would be absurd if that weren't so. It's true of all animals.

Nevertheless, it's just as true that many of our greatest intellects have often declared themselves mystified by the mind and how it governs our behavior – and continue to do so. There is a constant stream of scientific papers announcing some new discovery or hypothesis about our mind and how it works. It really is extremely complicated and its finished products – behavior – are rarely as straightforward as they are intended to appear. We get up, shave, get dressed, drive to work, drink coffee, come home, have a bath, watch TV, go to bed. We take all that for granted; yet just a moment ago in evolutionary time we didn't behave in any of those ways, even though the genetic makeup of our brains can't possibly have changed in such a short time. In other words the basic operations of the brain and mind must be more or less the same now as when we ran about naked, unwashed, unshaved, throwing spears at each other, dancing in the moonlight and casting spells on one another – no doubt taking it all for granted. It was this sort of consideration that made Freud look beyond manifest behavior to find where it all began.

1.1   The unconscious mind

That part of our mind is unconscious had been noticed by thinking people since Aristotle's day, but its fundamental role was not realized. Freud made it his business to unravel the manner in which it functions. He saw that the unconscious is the original basic mind that we inherited from our earliest ancestors. It's where all behavior of all animals begins, and in simple, so-called primitive animals where it ends. It's the repository of all our instincts, the main ones of which are many millions of years old. An instinct is a behavior determined by genes, triggered by a stimulus – from inside the animal or outside it – which the animal must, absolutely must perform. A triggered instinct can only be thwarted by the stimulation of an even stronger instinct capable of overriding it. The sensation of hunger, for example, triggers the instinct to eat. The stronger the hunger, the stronger the drive to eat. A hungry animal's instinct to eat, say, is almost instantaneously over-ridden by the sight of a predator hunting it. The instinct to flee can in fact over-ride all other instincts, for obvious reasons. Some instincts, like hunger or copulation, are temporarily extinguished by satiation. Others, like vigilance or aggression, are never extinguished (except when unconscious), though their intensity is raised or lowered by the volume knobs of contextual stimuli. In a safe place surrounded by kith and kin the intensity of vigilance is turned down low. Suddenly confronted by someone intent on stealing something from you the intensity of aggression is turned up high. These variable expressions of instinct can readily be observed in baboons, chimps, or ourselves, as would be expected, since they are driven by the same genes in all three. The identical, self-same genes. That is the all-important point. I will come back to this matter later.

For example, a hypnotized smoker (who wants to give up smoking) told that they will find cigarettes repulsive from now on does indeed react that way after recovering normal consciousness – but is unaware of the suggestion. While this clearly shows the unconscious working incognito on the conscious, it's a special case of the unperceived unconscious mind at work. Nevertheless, the phenomenon is general, and precisely because it's unperceived is understandably hard for many people to grasp. It constitutes the core of Freud's many insights into the mind and is the first step that one must take if the rest of our mind's activities are to be made sense of.

Freud was familiar with the Darwin-Wallace theory of evolution by natural selection and knew that there must be a functional reason for the covert nature of the unconscious. He saw that it was a mental device to protect the conscious – and thus the individual - from possibly dangerous instinctive urges, or frightening memories that, if known or remembered might impede the conscious' function. For the conscious is the means whereby the individual decides to act in the way that best serves that person's fitness – fitness in the evolutionary imperatives of fitness to stay **safe**, find **food** and **reproduce**. It must avoid influences that might impair its ability to do these things in the most effective way, which, as often as not, is to act extremely quickly. During our evolution in the wild, the conscious absolutely had to stay vigilant, ready to act without mental hesitation or physical delay.

It's vital to realize that our conscious mind does not *initiate* behavior; it only allows or disallows it. We do not consciously *invent* any of our behavior. It's all genetically stimulated in accordance with our ages-old, genetically inflexible instincts, which reside in our unconscious mind. What we *can* do, far, far more than any other animal, is consciously bend, stretch, redirect, restrain and otherwise mold our instinctive behavior in a complex variety of ways to better fit the reality of the moment. We can call upon the past to model and predict the future. We can, far more than other animals, empathize with others and by knowing their thoughts adjust ours accordingly. We can abstract ourselves from day-to-day external reality and think through purely mental realities. We can fantasize virtual realities and, rather unfortunately, replace palpable reality with delusions that are given the same (or even more) substance than actual reality. Some even think we can exercise a degree of freewill wholly independent of our instincts, beliefs, conditioning and learning – but that is moot and not essential to this enquiry.

The instinctive content of the unconscious is easy enough to come to grips with. There are several so-called primary instincts (or drives as many prefer to say), master instincts that are expressed almost continuously. Above all, we instinctively want to belong to our species, and specifically to our group or society. Within society, we instinctively associate with our kith and kin (where kith means 'known' and kin means relatives). That is the basis of what we call our identity, and is unswerving. It's our primary point of reference for everything else. The preservation of this identity is tied up with the more specific primary instincts of staying safe, aggression, eating, copulating (euphemistically called sex for humans and mating for other animals), which evolved to ensure that individuals maximize their chances of staying alive, competing, feeding and reproducing. The all-pervading aggressive instinct may dominate every other urge, or be applied like a turbocharger to boost the other primary instincts sufficiently to out-do any other individual homing in on the same object or goal.

I would emphasize here a fundamental fact of Nature for those who may not be fully aware of it: competition. All animals compete for a slice of the environmental cake, but for the most part this level, the species level of competition, is not between individuals, so that members of different species don't often fight each other. Competition at the level of hand-to-hand combat is primarily among members of the same species, and can be intense. Competition between species is insignificant compared to that among members of the same species. That is because animals only evolve into different species because no one species can fully exploit the entire environment. That would require an animal that is small as well as big, arboreal as well as terrestrial; aquatic as well as land living, etc. (This leads some to say there's only one species of animal with many different populations.) A species comes into being because random gene mutations gradually enable members of an existing species to exploit a range of foods or other features of the environment that neither its parent species nor any other can do as effectively, or is already adapted to. The parent species may remain alongside its newly evolved relative or go extinct, outdone by its descendant. Inevitably, the members of the new species must compete with each other, because they all depend on the identical part (or niche as some like to call it) of the environment. There are two reasons they compete rather than all live together in some sort of peaceable kingdom. One is that their genes keep undergoing selection for advantage, thereby maximizing the individual's ability to outdo its fellows – all of whom are going for exactly the same elements of the environment. The other, related reason is to keep pace with changes to the environment, primarily climatic, that are happening all the time.

This imperative of competition in Nature is extremely important because we have broken with conventional evolutionary competition in a way that has had the profoundest consequences for us. I will return to this in Part 3.

1.2   The Conscious Mind

The conscious mind is harder to come to grips with than the unconscious because it includes our perception of self, and the capacity to choose from a range of possible reactions to a stimulus. I don't mean to imply that the unconscious is simple; it isn't. But, the unconscious is the repository of our most ancient instinctive behavior that operates in a much more fixed, deterministic way than the conscious.

It needs to be said at this point that vast amounts of twaddle have been (and continue to be) written about this component of our mental apparatus – our perception of self. Just settling on what is meant by self excites people to all manner of sophistry and worse, most of which is in my opinion pure moonshine. The hokum all seems to boil down to two things: the persistence of the old concept of duality; and the intrinsic limitation to our powers of conception. The notion of duality is that our mind has a palpable, worldly, behavioral component, plus a separate 'soul'. This soul, treasured by god-believers and others infatuated by notions of spirituality, is above tawdry, worldly things and is where man's quintessential 'being' resides; all those things he aspires to that have, or at any rate seem to have 'raised' him above the brute animal. This unfortunate matter has profoundly affected our understanding of ourselves that has largely exceeded its use-by date.

The other problem- of conceptualization - is that we have not evolved the ability to perceive everything around us. Some things are beyond our natural equipment. Energy for example. We know it's there, even though it's invisible. We can measure the effects of energy as it travels through a system. We know (at least physicists claim to know) how much of it there is in, say, an atom. We can 'see' how it's integral to all processes – nothing can happen that is not mediated by a transfer of energy. But, what the stuff is we cannot say, quite likely never will.

When it comes to defining consciousness, to positing a testable theory of it, there may be a real problem of conceptualizing it; or it may simply be that we haven't yet learnt enough about the brain's electro-chemical processes. As a zoologist, I have no doubt that we will eventually be able to describe consciousness in electro-chemical terms.

> Man's desires are limited by his perceptions; none can desire what he has not perceived.
> **Blake**

Those who study brain function include some extremely ingenious and perceptive people who are already looking at the fine detail of the brain tissues involved; they are bound to come up with ever-deeper insights. For the time being we can't go much further than identifying the multitude of perceptions that constitute the 'body' of consciousness, and what the brain does with them.

For all the sheer flapdoodle about the mind science has already assembled a great deal of hard fact about the conscious that makes for truly fascinating reading. It's the final mental filter of our ideas and ideals that yields the best of civilization, the director and editor so to speak of our most penetrating, subtle and creative thoughts. Not surprisingly, it's extremely complicated and I could not possibly do justice to a comprehensive description. I shall only deal with those aspects that concern this enquiry without suggesting that I have more than touched on its intricacies. Fortunately, these intricacies relate to what the individual is capable of when acting in intellectual isolation from society; I'm looking at what happens in society, which is a lot less subtle.

To return to the definition of the conscious. I see no need to treat this as something intractable, as long as we are content to accept that we are apes (which of course even many scientists refute – but more on that later). The difficulty is only one of choosing a suitable model of the brain's functional structure, such as is known of it. One can define consciousness as the perception of one's presence in the world. The world comprises one's ego[2] or self, the perceptible egos (actual and imaginary) of others in our domain (i.e., those of other animals, both real and supernatural) and the environment. This is a compound perception, a 'reflection' of multiple perceptions supplied by sense organs, including the brain itself. The perceiver's ego is the master-perception, the sense of self; all other inputs are condensed and reshaped as they come in and compared with the master-perception. Both matches and non-matches (i.e., expectations and surprises) attract the ego's attention and trigger the next motor outputs.

The collective master-perception is self-evidently highly detailed. The detail is compounded from birth onwards, and continuously re-sorted, filtered, condensed and modified throughout life. We cannot remember everything we perceive in life – even in one day or hour of life – because we are bombarded with a vast and continuous stream of perceptions, the majority of which are banal and useless and immediately erased. Others might have been relevant at the time, and memorized, but as the relevance fades they become increasingly difficult to recall and eventually decay. Certain perceptions are strongly memorized and easily recalled throughout life, not always for obvious reasons or in some cases for no known reason. Most though are obvious enough – an alphabet, a tradition, an accident, a lover, a place, etc.

Each new detail, after its initial perception and condensation, is either immediately forgotten or added to the master perception and stored in memory, leaving most of the brain function free to attend to the next perceived deviations from it, which come in an almost constant stream. While asleep, it's (unconsciously) fine-tuned by condensation, by either wiping trivial short-term memories or retaining important ones, and resumes refreshed at every awakening. Condensation results in a ranking of significance, what we call conceptualizing.

The really important part about the conscious requires a mental effort to accept, because it's counter-intuitive. It's that the conscious **_does not generate_**

---

[2] The word ego has become synonymous with the plainer word self, owing to Freud's translator. Freud actually used the German 'ich', literally 'I', not ego.

our behavior in response to our received perceptions. Our responses are produced unconsciously. The conscious monitors those impending responses, *reflects* them is one way of putting it, so that we are instantly aware of what we are urged to do. The reflection into consciousness makes it *seem* as if we consciously generate those responses; but in fact they originate as gene-controlled reactions to whatever instinct is stimulated by the perception, or perceptions, of the moment. What the conscious does do is *vary* the reaction(s) it perceives as about to take place.

There is a complication, though. Somewhere along the evolutionary path we added another function to the conscious, one that differentiated us from cousin chimp, and all other animals. This addition was the capacity to project our self-awareness into imaginary scenarios, to create a virtual reality in the mind. By extending the consequences of an action into possible outcomes beyond the moment we expanded the range of choices of action. This capacity for abstraction was one of the most influential evolutionary adaptations Nature provided us with for it opened the evolutionary window to several other developments.

I would mention that all higher animals have unconscious and conscious minds. I find it astonishing that even some scientists think we are the only conscious beings. All animals can be rendered unconscious just as we can by sleep, trauma or drugs. A conscious state of mind is obviously evolutionarily ancient and essential to the coordinated existence of the individual animal. All of us animals are, when unconscious, physically paralyzed (except the autonomic system – heartbeat, breathing, digestion and other involuntary functions) and mentally reduced to the equivalent of paralysis. The basics are common to us all.

I will leave further discussion on these matters, particularly the third major element of the mind – our conscience - until Chapter 7 and deal first with the main features of how we think and act. There are a number of principles that shape all our mental behavior – conscious or unconscious – that Freud saw as essential to understanding our behavior. Allusions to all these principles had been made by many long before Freud; he was simply the first to describe them all systematically and show how they combine to give our thinking certain, predictable characteristics.

1.3   Antithetical nature of things

Freud emphasized the antithetical nature of much of mental function. The binary operation of stimulation and inhibition that underlies so much of our behavioral function. It occurs throughout language in that most words have antonyms, that we convey the bulk of information as either positive or negative. This reflects the simple fact that all stimuli are converted into perceptions by the brain, which in turn must be reacted to. To generate a perception that has no effect is a pointless operation of the mind and can hardly have been selected for in evolution. The initial reaction to a perception is ranked as pleasing or repellant, as positive or negative in some way. We make ourselves aware of this first ranking in the form of an emotion. These emotions range from mild to strong (itself an example of the principle of antithesis) and provide us with the basic criteria for our resultant behavior. All emotional expressions are antithetical pairs, which is really an NSDT statement, because one can only define an affect in the light of its opposite. If all animals were only hostile that would be the normal, neutral state not an optional state. We only recognize anger because the individual could be either calm or angry; two diametrically opposite affects. Thus fear or trust; loving or hating; nice or nasty; hostile or friendly; sad or happy; disgusting or appealing; surprising or predictable; guilt or innocence; anxiety or confidence; and so on.

The same goes for morals. There is no such independent standard as, say, good. Our perception of good, whether expressed or implied, is always in comparison to evil. If there was no perceived evil, then there could not be its opposite. Everything would simply be neutral – as indeed it is in Nature. Mother Nature doesn't care whether something happens this way or that. There's no hard evidence that other animals perceive moral rankings of any sort. It's only we who make such distinctions as part of the mechanism of our conscience. Our ego rates something good or evil (or its alter ego; god or the devil) simply as one of its many perceptions of what we want, and are pleased by, or its antithesis, what we don't want. If our conscious is to make a choice between two possible actions arising out of an instinctive urge it must have sound criteria – life itself may depend on the choice. Good or bad, right or wrong, tolerant or intolerant.

1.4   The Pleasure Principle

One of the most important of Freud's insights was that *everything* we do is designed to be satisfying, gratifying, pleasing – a defining characteristic he called the Pleasure Principle. A zoologist would say that to be valid this would have to be the same thing as maximizing evolutionary fitness. Obviously, we don't think in 'evolutionary fitness' terms. No one goes around supposing that if they get up early enough to be first in line for the Boxing Day sales that it will improve their evolutionary fitness. Getting there first results in a pleasantly triumphant emotion, nothing more. Freud clearly saw in the pleasure principle that the mind was using a simple and highly efficient device to keep the individual maximizing evolutionary fitness. According to that principle, we only act to maximize our perception of pleasure, whether it be immediate or postponed, physical or mental, direct or sublimated. By so doing we achieve what the genes for that principle were selected for – improved fitness. But, pleasure of itself is not the ultimate goal. It's just the emotional signal to the conscious (i.e., the self) of a successful action. From the evolutionary perspective, pleasure only indexes fitness.

Of course, pleasure itself had to evolve first – but there's no need to elaborate on such ancient genes that we already had when our ancestors were frogs, and before. If you think about it, the pleasure principle is really an NSDT[3] statement, because it's absurd to suppose that the mind would ever prefer the unpleasant. You may argue that on waking up late with a hand-grenade of a hangover choosing nevertheless to go to work is an example of choosing the unpleasant. But, that's too simplistic; the choice takes into account the future consequences of missing work.

This raises an important feature of the mind and how it works. Freud emphasized that pleasure once achieved 'settles' the mind so to speak. It's as if a build-up of mental energy has been discharged and the mind is now free to pursue the next objective. Displeasure is the reverse; the mind builds up energy trying to avoid displeasure, which will continue until the wished-for pleasure is achieved. This has an important consequence. The avoidance of displeasure can take a lot of time and mental effort, and in the contemporary world can in fact often be what many spend the greater part of their time doing, so that it can seem as if the pleasure principle doesn't hold.

The pleasure principle has partially seeped into contemporary society's awareness, along with the existence of the unconscious and some of its influences. The kinds of unconscious thoughts that underlie wish fulfillment, Freudian slips (doing or saying things 'by mistake on purpose'), jokes and so on

---

[3]No Shit Dick Tracy, NSDT. The patently obvious. From Chester Gould's laconic comic strip sleuth, Dick Tracy. As in, 'The blonde lay in a pool of blood. There was a .45 caliber bullet hole in her forehead. The blonde was dead.'

are widely understood nowadays, as they always were to many creative people. The French expression *tuer son mandarin* refers to the idea that if someone in Paris, certain of their anonymity, could wish a Chinese mandarin dead and thereby become exceedingly rich they would indeed do so unhesitatingly. That we act to please ourselves is common knowledge. The problem with these mental functions, in particular the pleasure principle, though, is that we don't always see it operating. What Freud realized is that it *always* operates; it's the criterion by which the conscious makes *every* decision – not just the obvious ones, and not just the perceived ones. Specially not just the perceived ones. For example, a politician who breaks with the conventional political antipathy to reducing reliance on fossil fuels might announce that he will fight to reduce fossil fuel use in the interest of combating climate change. On the face of it, he is seen to be acting in the wider public interest. In reality he is driven by self-interest, a wish perhaps to gain attention, or to curry favor with his electorate. Politics is a good place to see the pleasure principle operating as so much of what politicians claim is transparently self-seeking.

We have to remember when trying to unravel the operations of the mind that it's extremely difficult, and obviously so, to accept that so much of our behavior is unconsciously motivated and therefore unperceived. What we cannot see is in the first instance dismissed as non-existent. What the eye doesn't see the heart doesn't grieve over. NSDT. Naturally. When we consciously decide to act in a particular way that does not appear to have any unconscious motivation we assume, equally naturally, that it's entirely a conscious decision completely independent of any unconscious drive to seek personal pleasure. Therein lies the single biggest problem faced by civilization - the fallacy that we and specially our leaders can and do make decisions free of personal gratification for the altruistic good of society. We all of us speak constantly of the public interest, of policies and actions designed for the improvement of society, the honor of the nation and the advancement of civilization. If only. Looked at objectively in terms of human behavior there's little that stands up to the high-minded ideals claimed. The reality is that it's easy to tease out the real, self-serving reason for every action anybody takes. Such actions may also result in something that serves the public interest, but that is rarely the motivation behind the action however strongly the actor may think it is.

Leaders are after power, control, and mastery. The led are after good stuff, a society that keeps them safe, well fed, healthy and free to gambol about as they please. The leaders concede, unwillingly, that to keep power they have to amuse the led sufficiently that they do not turn to other leaders who promise to please them more. The led concede, unwillingly, that they have to obey the leaders' rules sufficiently that they do not get too harsh. That sums up how things are under contemporary efforts at democracy that include the rule of law. Dictatorships dispense with the irritation of public opinion and have only to concentrate on scotching the next coup.

To understand how we fool ourselves and others into thinking we are behaving as we would like to behave or how we want others to believe we are behaving, it's necessary to be aware of the more important devices and attributes of our behavior. That this results in something of a stodgy list is unfortunate, but hard to avoid.

1.5   Defense Mechanisms

An important concept introduced by Freud was that the mind has evolved a number of instinctive 'defense mechanisms' – mental devices that evolved to mitigate the conflicts and anxiety caused by instinctive urges where these threatened to overcome the cultural restraints that we humans put on all our instincts. For him, the most important was what he termed *repression*, whereby in childhood a traumatic experience that might impair a child's maturation is subconsciously repressed, i.e., kept unconscious. This is an ancient instinctive process that does not operate in adulthood, when trauma has to be dealt with consciously. The concept of repression is tendentious and will remain so until a great deal more brain function is unraveled. The Freudian hypothesis revolves around incest and the obscure mechanisms that evolved to avoid it in a highly social species such as us. Incest is undesirable because it tends to reduce genetic variety. In social animals it occurs less often than chance predicts. That means that behavioral devices have evolved to avoid it. Chimps rarely copulate with their mothers as the result of complex patterns of copulatory behavior in both males and females. Combating incest in us humans was complicated by the evolution of constant copulation for pleasure, which inevitably meant that family members would perceive each other as copulatory opportunities; daughters for fathers, sons for mothers. Freud observed that this was further complicated in early childhood by a reciprocal attraction of fathers for daughters and mothers for sons - a phenomenon he termed the oedipal conflict. He interpreted repression as a device by which children archived the memories of incestual events or temptations in the unconscious. Freud considered that a failure of repression in childhood was the root cause of much neurosis and the pathological condition of pedophilia. As I say, the whole topic is contentious involving as it does instincts and cultural taboos that many find extremely unpalatable and difficult to deal with. I shall say no more about it, intriguing as the matter is, because it does not affect this study.

Lying   For me by far the most pervasive of all defense mechanisms is lying. It seems paradoxical at first that society condemns lying as wrong yet accepts it in so many ways. There is no law against lying; in fact, the law condones lying. In every contested legal case, by definition, either the prosecution or the defense is lying. The vast majority of convicted criminals deny their crimes. The law does not punish them for lying, and most people are unmoved by the fact of their lying. The law only forbids specific kinds of lie; perjury, criminal fraud, plagiarism or libel where a lie is deliberately made to injure or steal from another person. Politicians are not punished for lying – only for misleading parliament. Misleading is lying, but is a euphemistic softening of it, and rarely punished even though it's an almost daily occurrence in parliament. Since they are all liars from time to time it's always going to be risky for the pot to call the kettle black. Moreover, there's no rule or law which forbids politicians lying to their electors; only to each other, which is self-evidently self-defeating. Yet lying to their electors, to whom they are putatively responsible, is ethically far worse than lying to each other.

The purely mental composition of lying is why it generally escapes the law, which, for the most part, deals only with material things; theft, violence, contracts, illegal acts. And this points to the reason why lying is so pervasive and so resistant to the ethical protests of our conscience. We can be sure that lying is an ancient instinct and not at all a recent product of our conscious minds. We can be sure because cousin chimp is a liar, a master of deceit and I have no doubt the genes for this are among those we share. In other words, the use of deception is an ancient adaptation that must have had strong selective value or it would have disappeared long ago. Cousin chimp has been observed to skillfully deceive to copulate, get food, eat food secretly, to stockpile and hide missiles to throw at unsuspecting zoo visitors and to avoid being hurt by higher ranking chimps. Monkeys and other animals have been observed using deceit similarly. It's adaptive because it works to the individual's advantage. More importantly, though, we routinely employ deception on ourselves, long before we employ it on others. As Freud pointed out, the evolution of the conscious management of our nagging instincts involved mechanisms such as repression, sublimation, transference, etc., all of which can be seen as distortions of plain instinctive wishes, urges and drives that nearly always work by rationalization or miss-representation of the instinct involved. In other words, by deception. We fool ourselves in all sorts of ways to conceal forbidden or violent instinctive proddings and make them appear rational. If you love irony as I do you cannot but be struck by the irony that lying is normal, natural and essential if we are to be civilized.

It all goes back to the primacy of the Pleasure Principle that we are genetically bound to do only that which benefits ourselves, and where we do something that benefits others, it's only because it primarily pleases us to do it, irrespective of the rational, apparently altruistic benefit. Yet we usually vehemently deny any self-interest in so-called altruistic actions, both to ourselves and others. Such self-deception is simply one of the many artifices of the mind to protect or reinforce the ego. Such deception is unconscious when used defensively, but conscious when used offensively (in the sense of deliberately).

It follows that we must be ready to employ artifice, deception, cunning, bullshit, humbug, bluff, lies, etc., without hesitation. The instinct to stay alive is the most powerful of them all, and is what is meant by the psychoanalytic assertion of the Pleasure Principle, that we only act to please ourselves, directly or indirectly. And because we are social animals, we are constantly competing for advantage within the group (as well as with other species). Mayr's competitive exclusion principle makes it clear that if lying works (and for sure it does) then it will be used.

> Honesty is praised and starves.
> *Juvenal*

There is a class of things we get by force, by the instinct for aggression, and brute force is about the only truly honest thing we do. A few other things we get honestly, by luck or because there's no need to dress them up in any way, but most things have to be competed for. And competition always involves some degree of bluff (i.e., deception) except where it manifests as plain violence. As far as the limited evidence goes lying has always been condoned. The so-called Christian Ten Commandments are an example. These were simply hi-jacked by the monotheistic prophets from an ancient set of social rules that were already well-established by the time writing evolved (the claim that the religion in question established its own ten commandments is itself a lie). The ancient Egyptians for example when they began writing around 5,000 years ago listed 30 rules that society was supposed to abide by. Most concerned specific kinds of theft – which appears to have been the commonest crime in their society, as indeed it is in ours. They also categorically forbade killing, adultery, sorcery and plotting. Only one alluded to lying, and it wasn't general, but specifically forbade the perversion of justice by getting witnesses to lie (or bearing false witness yourself). As such, it was an adjunct to whatever was being tried, rather than an issue in itself. It's really an NSDT matter because all theft only comes into juridical existence when someone accuses another of it; perjury simply had to be forbidden. The Egyptians already had a well-developed justice system that had whittled down the undesirable things that people commonly did into the basic set of rules that still underlie religious commandments and secular law. No one ever saw reason to add lying generally to the '10 Commandments,' and these basic rules of conduct have stood unchanged for uncountable millennia.

I find this intriguing because the search for truth has been the ultimate goal of philosophy and intellectual activity generally ever since written records began, and presumably long before. Religions always claim it. It's the defining criterion of science. Its antithesis, lying, in all its forms, is universally deplored, but only children are punished for it. Once adult there's no formal punishment for lying (except as a court witness), although in certain contexts it's used to discredit individuals.

The answer to the question as to why we simultaneously deplore lying yet condone it is elusive. There's always a problem with things that are not physical acts. Murder, adultery and theft are tangible, provable losses of some form of property, but lying, even when provable, does not leave a physical trace in the form of something material the victim has been deprived of. When a public figure lies he doesn't steal from or injure us or society by his lie, nor does he lose much by it (he knows as do all politicians that his electors are only interested in what he can get for them, and they don't care how). Insofar as the law is concerned to redress wrongs, it needs something material to form the basis of the punishment or compensation.

To me the basic flaw in democracy (the word originates from the antonym to aristocracy) is that it replaces the tyranny of the aristocracy or the dictator with the tyranny of the majority (aka the twittering mob). For those of us who deplore the simplistic, bodily-driven opinions and desires of the twitterati, democracy simply means we can never be ruled by people of our own choosing. This is a depressing irony to say the least. It's also a fact that the vast majority of Western democratic elections are won with comparatively small majorities, which means that nearly half the people will always fail to get the representative government democracy is supposed to guarantee them.

On top of that we are not free to choose our rulers (as democracy mendaciously claims) because we are forced by law to elect only a small class of people –party politicians - who themselves are not free to act according to conscience (assuming some of them possess one), but only according to the wishes of the mob who themselves are not capable of exercising conscience over greed, prejudice or self-interest.

Democracy, like religion, is an illusion, i.e., a lie. And its mechanism is the rule of law, which is only indirect force. Yet democracy seeks to promote the principle of freedom. And this rule of law does not provide for any device whereby rational thought (in government) must prevail over emotion, i.e., instinct still reigns supreme. And on top of all that the tyranny of the lie of religion still dictates much of what the mob and its politicians do, notwithstanding the illusion (the lie) that Western governments are secular.

Lies, lies and more lies. The individual governs his ego with considerable distortion of instinctive truth; the group governs itself by employing (inter alia) the whole gamut of lies in every aspect of civilization. These things haven't changed much since civilization began and probably won't for a very long time to come.

Denial   After lying, perhaps the most common defense mechanism that is relevant in the present context is denial whereby one denies the existence or truth of an unwelcome thought, prejudice, urge or event. It's strongly expressed in the categorical dismissal of, for example, a violent urge, the Holocaust, evolution, climate change or some such as mere fictions. Denial does not involve any interpretation of evidence; just outright rejection. It's more weakly expressed in for example 'I'm not racist', 'I never said that, 'That's not what I mean, 'I don't want to be rich, 'and similar. We use it all the time usually without noticing because it's an instinct; it's the unconscious mind that activates it.

One of the most conspicuous examples of denial is the subject of this book – the unconscious denial by politicians and others of the glaringly obvious threats to civilization posed by the blind pursuit of economic expansion and population growth.

Projection   Another important defense that is frequently used by all of us is *projection.* Like the others it's instinctive and for that reason is always unconsciously initiated, even though careful thinkers may make themselves aware of it. It refers to the unconscious projection of one's own wishes, anxieties, thoughts or prejudices into someone else's mind. For example, a woman might accuse another of being a sticky-beak, when in reality it's the woman herself who is a nosey parker, but can't admit it or doesn't even realize it. Shifting the blame. The pot calling the kettle black. 'She's just envious of me' when in reality the accuser wishes the other *was* envious. 'I'm not trying to interfere – I'm only making a suggestion.' '*You* should've seen it was going to rain' for *I* should've ....

Another common projection is the apparent sympathy that arises out of our capacity for empathy. When something unpleasant happens to someone we can empathies with that person and feel sympathy for them. What people seldom realize is that the feeling of sympathy arises from the projection of one's own feelings on to the victim. In fact, we often give the game away when for example we see someone killed or injured in, say, a car crash. 'Oh wow – that could've been me.' It's common among conservationists and animal lovers to express pity for an injured or abused animal. But, the pity stems from one's projection of oneself into the incident. The animal has no concept of pity or cruelty or misfortune – these are all sentiments we possess about ourselves. Feeling them for animals is to soothe oneself. I don't wish to scorn such sentiments, only to point out that their evolution is from our own feelings about ourselves. Empathy evolved as a sophisticated device for protecting and strengthening one's ego while at the same time strengthening group bonds.

Paranoia is an extreme form of projection in which we project our own guilt and anxieties onto other, often imaginary people. We feel they are accusing us of what in reality we accuse ourselves of. It can become pathological in severe cases.

Rationalization　　　　A universal defense against the reality of personal wishing that virtually everyone employs almost continuously is rationalization. It means the hiding of an instinctively driven urge or wish behind an allegedly utilitarian, objective or altruistic purpose. 'I joined the air force to serve my country.' Really? Or was it more that flying hot airplanes and blowing enemy targets to smithereens appealed strongly to something deep inside? 'I wear a low-cut dress because it's fashionable, and not as you allege to encourage men to look at my boobs.' Or, 'I don't care whether or not people read my books– I write them only for my personal satisfaction.' Sure thing. 'I married him because I love him, not for his money.' 'Government taxes you for the public good.' A compelling rationalization that relies on the partial truth of the outcome. But, taxation began as a king plundering his subjects' labor to fuel his grip on power and continues to be the underlying motivation of all politicians. Dictators illustrate this perfectly, and all politicians, as so many have so often observed, would be dictators if they could.

A common rationalization among those who go fishing or shooting is that it's okay to kill a wild animal as long as one eats it. It would simply be a wish to kill if one did not eat it. That such people do not need to eat it because they have plenty of food is never seen as giving the lie to the rationalization.

These examples are relatively trivial, everyday rationalizations that many people are well aware of, even though they continue making them. What many do not seem to appreciate, though, is that exactly the same mechanism is employed by our leaders (who definitely don't seem to get it). When a leader invades another country, for example, it is always justified by an alleged need to combat terrorism, or to restore peace, or to replace a dictator with a democracy or some such grand objective. These are easy rationalizations to make because most people regard them as justifiable or a 'good thing'. Yet I (and many others) find this naïve in the extreme. Leaders are instinctively driven by the form of aggression that manifests as the pursuit of power. It's the source of their energy and determination. We can be sure that the underlying motive in the example is the wielding of power. It would be a rare leader who could objectively set aside his drive to power and sublimate its energy into a truly rational, dispassionate objective.

Regression   A rather specific defense against aggressive or violent tendencies in oneself or the world at large is regression. In this one regresses to a more childish, essentially innocent, perception to protect oneself from one's own capacity for aggression or other expressions of violence or callousness. In my book *The Gardeners of Eden* I showed how the conviction of the need for the conservation of Nature is a classic example of emotional regression; a retreat from the intolerable. It could be coldly rational, but it isn't – the vast majority of conservationists are extremely emotional about their cause.

1.6   Idealization

An important, purely mental instinct is the drive to shape wishes into ideals. We are not born with social objectives, at least not abstract ones that do not operate until adulthood; but we obviously have a template for cultural goals that represent a coalescence of several cultural desires. This is a difficult aspect of the mind because it's neither a plain instinctual drive nor a simple cultural molding of physical behavior. It's purely intellectual and our current knowledge is insufficient to analyze such brain functions in electro-chemical terms. We can only proceed, as did Freud, with the observed behavior. He emphasized the importance of the fact (that had been remarked upon by many before him) that we culturally establish, and perpetuate, idealized, essentially unattainable goals. These are comprehensive, widely endorsed goals that encapsulate society's strongest wishes. They include the ancient, classic ideals of peace, love, freedom, tolerance, equality, justice, cooperation and democracy. It will be immediately obvious that regardless of how many times we claim to be pursuing these ideals (and they have been repeated ever since the first thinkers we know of wrote them down) in practise society tends to their antitheses – to war, hatred, mastery, intolerance, hierarchies of race, rank and wealth, injustice, coercion and dictatorship. It's hard for many to accept that the latter are the primal tendencies. The former are the idealistic antitheses, mental artifices rather than mental realities. No one has to campaign for inequality, or racism, for example. But, we certainly do have to campaign for equality or tolerance. The one is spontaneous; the antithesis is contrived.

What we are compelled to do, what we are instinctively urged to do, is not necessarily what we think we are doing. Our intentions as we consciously sense them are often those that have been idealized to fit goals that strengthen the group rather than its individuals. That is the function of the mental process of 'cleaning up' raw instinct.

The function of ideals, the selective pressure on them, is to make the individual keep striving to fit better into its current environment. Since the group is an assembly of individuals group ideals must not be at the expense of individual satisfactions; such ideals must improve the chances of both. Thus if individuals are killed or crippled in pursuit of group ideals such losses must not exceed the group's net gains (over the long-term; high casualties over the short-term might be justified, as for example in war between groups).

Freud called this masking of instinct by cultural ideals a façade, the façade of civilization. Like most façades, it's easily torn down. Whereas we cannot destroy our instincts, we can – and do – readily destroy the cultural artifices devised to mask those instincts. Sport is a good example. The drive to fight each other for dominance, to outdo one's competitor, has been culturally reshaped in the form of sporting contests that hide the intrinsic violence of physical competition under symbolic rituals of procedure – the rules of play, contact, scoring, etc. This ritualized violent competition has been pursued by all civilizations and is a prominent feature of society. What we tend to forget though is how often the rules of ritual are broken by the instinctive urges of players, and how essential it's that the rituals are *constantly* policed. The tendency to regress into actual fighting is always there and often gets the better of the ideal, as everyone knows. The urge to cheat is also always there and clearly impossible not to succumb to.

The wide-ranging ideals of society are not the product of society itself but of individuals within it. And the true origin of our collective ideals is the idealization by all of us of our egos, of our perception of ourselves. Our egos are driven by our instincts, which we 'domesticate' by all the devices mentioned above to create an idealized version of our real selves. We claim all the time to ourselves and others to be pursuing all sorts of altruistic or sanitized goals that are really all ideals that 'civilize' our brute instincts. Reality of course is constantly butting in and sullying the purity of the ideals.

Sublimation.   Usually classified as one of the defense mechanisms, sublimation can also be seen as a form of idealization. It describes the displacement or redirection of aggression and sexuality into socially acceptable behavior. It's part and parcel of creativity and is commonly seen in those who stand out in art and science as well as in sport and entertainment or other socially admirable endeavors. It represents a combination of unconscious as well as conscious efforts to modify raw instinct and is characteristic of energetic individuals.

Other Features of the Freudian Model

The Passionate Genes, the Super Powers of the human psyche

Aggression and Copulation

I have singled out these two elements of the Freudian model because of their strength and ubiquity in our behavior. In fact, it's hardly I who singled them out — they do that themselves. No one would doubt that in Western society these two instincts account for the greater part of our attention, even though great advances in society's attitude to the copulatory instinct have taken place in recent decades. In Freud's day, acknowledgement of one's extremely powerful urge to copulate and discussion of it was strongly prohibited by society and was therefore deeply imbedded in people's super-egos. It was the observation of this fact arising out of his clinical work that led Freud to conclude that this instinct and it's frequently unsuccessful cultural control was behind most of our behavior, including aggressive actions. It was only late in life that he recognized the independent existence of aggressive instincts. While the consequences of this, to my mind, misdirected some of Freud's interpretations, his dismantling of the resistance of individuals and society to the open acknowledgement of all things sexual ranks among his greatest contributions.

It's interesting to note that this Western demonization of copulation didn't happen in the parallel eastern civilizations. Such things as the Karma Sutra and the Thousand and One Nights were in Freud's day pornography — an attitude quite foreign to the East. I have no doubt that the reason was the determination of monotheism to outlaw the indulgence of pleasurable instincts as marks of sin. Of course, the deeper reason was the exercise by the priests of power — you can't have people fooling about amusing themselves. They must feel guilty and inferior. Power needs discipline and people must be made to feel guilty about their undisciplined urges to satisfy their barbaric habits.

1.7   Aggression

There is an extraordinarily widespread prejudice about aggression that needs to be scuppered at the outset, and that is the notion that aggression means violence; that the two are synonymous. Even some scientists fall foul of this mistake. Violence is certainly an expression of the instinct for aggression – but it's far from being the only one. All violence is aggression; not all aggression is violent. I think the confusion arises because violence is so visible and so drastic, whereas the more everyday expressions of aggression are far less evocative. Aggression is the expression of those of our genes that push us to compete with one another in the raw Darwinian sense – it's really only another word for competition. It's a complicated instinct with a wide range of intensity from trying to be the first to grab something attractive that isn't performed so strongly as to result in physical violence, to violence so intense as to try to kill the opponent. Aggression is the formal name for what I've called our constant propensity for want, want, want; more, more, more. Statistically, theft is far and away our most common crime. It's the simple expression of the aggressive instinct to grab what you want by whatever means you can come up with. We may be reluctant to admit it. Who does not feel, however fleetingly, the urge to take something valuable that you come across by chance. The expensive sunglasses, the forgotten smart phone or umbrella, the wallet dropped on the pavement. The cultural suppression of the instinct can easily fail under such circumstances.

Along as aggression is seen as the expression of competition, it's a relatively prosaic adaptation that evolved in all organisms as a basic element of the suite of vital instincts. It almost goes without saying that by its very nature its evolution was accompanied by devices for restraining it. No organism can afford to be driven by raw aggression every time its attention is drawn to something it wants. The chances of success must always be weighed up and its intensity modulated according to circumstance.

In our case, a slow-moving biped no longer much good at climbing trees, whose muscular strength was fast diminishing, there had to be some adaptations that countered the concomitant vulnerabilities to predators. I suggest that the primary one was aggression. We became more and more aggressive towards anything in Nature that stood in our way, be it a lion, a rival, a reticent female or out-of-reach food. At the same time, selection strongly favored a cleverer brain that operated less and less on hard-wired instincts and more and more on conscious, cultural modulation of instinct, plus individual mental nimbleness.

A word about the association of predation with aggression. It has become a given that for much, perhaps all, and certainly our Pleistocene evolution we were predators. It's also assumed that hunting was predominantly done by men (as is the case with cousin chimp). Animal behaviorists have clearly shown that predation does not involve violent aggression as is displayed by most male animals among themselves, or the sort of violent aggression expressed by animals under attack. Predation is exclusively for food and does not necessarily call for more aggression than is normally employed by any animal in getting food. A lion pulling down a buffalo is no more aggressive than a giraffe picking leaves off a thorn tree before they turn toxic. There is no reason to suppose that we were any different as predators. Yet many including some scientists still confuse predation with violent aggression. When a shark catches and eats a swimmer the victim is invariably seen as having been attacked, with all the undertones that go with the notion of attack. It's partly a semantic consequence of there being no other word commonly used to describe the moment a predator tries to subdue a prey animal; but that itself is because people until recently never distinguished between attack and catching prey. I make the point because our predatory history is not, in itself, evidence of our aggressive instincts. Aggression is a visible manifestation of the instincts to compete with one another for access primarily to food and reproductive opportunities. Since the adoption of farming for food, competition has shifted onto land as the main prize.

This point is also relevant to cannibalism, which has joined incest and murder as morally forbidden. It's an NSDT statement that like all things forbidden there would be no need to prohibit it if the prior urge did not exist. As long as cannibalism was simply predation or scavenging it was not driven by aggression. Whether we hunted each other purely as predators or mainly as competitors for the same piece of turf is probably unknowable. Whatever, it appears that we readily hunted and ate each other. Many human bones found in archaeological sites show signs of having had the flesh or marrow removed. Such bones go back at least 600,000 years to Neanderthal and related sites. We cannot say what led to the owners ending up as the Sunday roast; whether they were killed for food or incidentally eaten following some other cause of death. That the meat was diligently scraped off the bone suggests that the eaters were either extremely hungry or rather partial to long pig. Probably both. For example, the remains of 13 cannibalized Neanderthals found in El Sidron cave in Spain all showed signs of lifelong nutritional stress. For some Neanderthals at least hunger was chronic and meat – long pig included - must have been highly prized. Cannibalism for food has been extensively documented and was widespread right up to contemporary times.

Among zoologists and other scientists, it's now a commonplace that we rank as an exceptionally aggressive animal, with men more so than women. In the 5000 years since writing began the extent of recorded pillage, rape, torture, murder, massacre and war is truly spectacular compared to all other animals, including cousin chimp. Adjusted for population size, cousin chimp is as violently aggressive as we are – which is hardly surprising given that we are genetically so alike. The only difference is that chimps aren't cruel whereas we are capable of any imaginable cruelty. Prior to writing, there's plenty of archaeological evidence of a similar state of affairs going back at least to the replacement of foraging by farming. Whether or not farming led to increased violence in the form of war over land and mastery of people we can't say, but to me it's probable that our enthusiasm for violence began evolving from the time that we began to assume the physical features that define us, i.e., from the beginning 19 million years ago.

The study of the evolution of aggression commonly uses the cost-benefit model, which allows a quantification of the various parameters that must be accounted for to determine the overall selection pressure that led to the genes for aggression. The benefits of employing aggression include the value, abundance and location of the food, reproductive opportunities and security being sought; the costs include the risk of injury or death, the time and energy investment and the suite of psychological damages that might result. It seems to be a truism that herbivores are far less aggressive among themselves than carnivores or frugivores (such as chimps). This is because herbivores exploit a widespread, freely available, non-perishable, leafy food that does not force individuals to congregate; whereas meat and fruit eaters depend on highly localized, briefly available foods that inevitably lead to concentrations of individuals and fighting over access.

The benefits of aggression are what animals fight for; the costs depend on how they fight. In our case, the aggression bias towards men is partly explained by sexual competition. Women do not need to fight over men because any man will do for reproduction. Men fight over women because potentially fertile women are scarce. Most eligible women are either pregnant or lactating and the rest are only fertile for short periods each month. All this is reflected in our sexual dimorphism, men being some 15 percent bigger and stronger than women. In mammals generally, sexual dimorphism only evolves in species subject to sexual selection. In solitary and monogamous species there's little if any sexual dimorphism and little sex difference in aggression levels. In species that group (such as us), there's potential for an aggressive male to monopolize more than one female and benefit thereby in increased reproductive opportunities.

The aggression bias towards men is also explained by our fighting over land and at least since farming over the mastery of people, though I suspect that mastery over people goes back much, much further than farming. Adjusted for scale, cousin chimp is the same. He patrols his territory with alliances of males, and murders (if he can) males from neighboring territories. He expands his territory (and therefore food availability) if he can and welcomes foreign females. The benefits of such behavior can be considerable as long as the costs in casualties are less than the 'enemy's.' I need hardly say how locked into this form of aggressive behavior we humans are. Dominant leaders gain mastery over people, build up their numbers and wealth until they believe they can overpower their neighbors, whip up a war, seize the land, rape the women and crow over their success. Over and over again, sometimes winning sometimes losing; but never losing the appetite (the instinct) for yet another try. That is one of the downsides of competition – it never ends. Complacency can only end in tears.

1.8　　　Copulation

Bipedalism - standing up all the time - dragged our noses away from our crutches and our original reliance on smell faded as selection favored visual and vocal communication. Copulation evolved for reproduction and in all mammals except us and cousin bonobo that is its sole function. The instinct to copulate is extremely powerful when a female is on heat (estrus), which is always signaled by odor with, in some species, color as well. At other times, there's no interest and neither sex shows emotional pleasure during copulation. Although our females still ovulate cyclically, the event is unperceived. Constant copulation evolved to ensure that undetectable ovulations continued to be fertilized. Something new had to stimulate the copulatory instinct once a particular smell was no longer detectable. The new stimulus that evolved was plain old pleasure. NSDT. Making copulation so pleasurable that it became an end in itself ensured that vital constancy. Only one other animal has evolved this unusual adaptation – our close cousin bonobo, who actually copulates even more than we do; and a lot less furtively.

Constant sex in both of us may also have been selected for another reason – to increase sociability between males and females. It was yet another device for reducing plain instinctual aggression as a mediating force in groups. Cousin chimp males who like all other mammals only copulate when a female is in estrus use violent aggression against females as readily as they do against males. Bonobo society is quite different. Group order is maintained by alliances of female friends, and male bonobos, though still bigger and stronger as in chimps are rarely violent to females.

I came across an estimate of the number of copulations in Britain that resulted in a fertilized egg (0.1% of the 900 million heterosexual copulations a year). I couldn't resist the temptation of extrapolating that to the global number of annual ejaculations and, by using the much-loved measure of quantity – Olympic swimming pools – estimate the amount of semen involved. Assuming the British estimate to index the global average, and after adjusting for masturbation and homosexual activity, I estimated that there are globally around 200 billion ejaculations per year. That is around 10 billion liters of semen, which would fill 210,000 Olympic swimming pools. I doubt that enough Olympic pools exist to confirm this.

There are other features of individual behavior that I won't go into because they throw little light on the points I've tried to emphasize: that by far the most influential principle behind our overt behavior is that it's self-serving, and typically subverted from its elemental instincts. We are social animals because the group is a sub-division so to speak, of our environment, occupation of which is advantageous to individual survival. Access to food and reproductive opportunities is accentuated by the reciprocity of group members, as is security. But, the individual must always satisfy his own, selfish interests first; the group is a feedback mechanism that only works if it's built on the individual success of a majority of its members. Adult chimps and elephants and lions function perfectly well living entirely alone. So can we. Group membership of all such species is primarily advantageous in survival and reproductive terms of the individual – not of the group. The group as such is not an evolutionary entity – merely a convenient assembly of individuals seeking the best they can for themselves.

Each individual has to make a mental model of everything conscious; nothing beyond the autonomic system comes pre-packaged in a fully operational form, though we do have genetically determined templates for setting up beliefs, relationships, language, morals and so on. These templates are generic so to speak and depend on the cultural inputs of the day and place for their completion. Without cultural input they are dormant; one isn't born speaking a language or believing in god. Just like other animals, we are born with the essential autonomic instincts, and the ancient, basic instincts; but not much else. Many of the things we do as adults that we take for granted have to be copied from those around us and installed into the blank templates in our brains. Selection has favored plasticity to enable a wide range of rapid – even immediate – adaptation to changing circumstances. The autonomic instincts in stark contrast must be impervious to short-term or outside influences. Breathing, heartbeat, or digestion have to do their thing regardless of whether the individual is fleeing a predator, giving birth or fast asleep.

****

My purpose in describing mental defenses was to emphasize as strongly as I could the primacy of the self, that everything we think and do is to boost the self and that this is exactly what must happen in behavioral terms when natural selection operates on genes. Because selection acts only on the individual – not the group or the species. This is really an NSDT statement because groups or species don't have genes and cannot therefore undergo natural selection. Mutations can occur that enable individuals to fit better into a group and gain survival and reproductive advantages as a result; but that is still an individual not a group adaptation.

To sum up individual behavior. We have an unconscious mind consisting largely of our evolved instincts, which are our gene-controlled mental responses to the bombardment of stimuli constantly emanating from the outside world and our own minds. Some instincts serve the autonomic system, which operates whether we are conscious or unconscious. The other instincts, the ones that concern us here, serve the conscious motor system, i.e., are only activated when we are conscious. That is because they are variously inhibited or enhanced or deflected by the conscious mind (though we are not always aware of it). No instinct ever sends its genetic stimulus to act in its raw, pure state[4]; it's invariably modified in some way (at least in normal individuals). The modification may follow a learned or conditioned sequence, or a culturally dictated sequence (i.e., directed by the conscience) with little or no conscious reflection (and therefore little or no awareness). Or the modification may involve a greater or lesser degree of conscious discipline or review.

Conscious review is the operative phrase. In considering what to do with instinctive promptings the conscious follows its own set of genetically controlled procedures, i.e., it does not 'make up its own mind' so to speak; it does not exercise 'free will.[5]These procedures follow one invariable rule; they obey the pleasure principle.

---

[4] It's possible to experimentally evade inhibition and evoke an instinct at 'full strength' by stimulating the appropriate part of the brain – short-circuiting it so to speak.
[5] Some believe we have a degree of free will. It's tendentious and if true it's hard to find evidence of it, or imagine the neural mechanism.

It remains to say that the foregoing describes the main features of our individual behavior as it concerns my purpose, but cannot by itself explain the whole of our relationship with Nature. That requires taking into account one more aspect of our basic behavior; how we behave when we form groups.

Chapter 2   Group Psychology

Nor should we listen to those who say 'the voice of the people is the voice of God, for the turbulence of the mob is always close to insanity.'
Alcuin to Charlemagne, 1200 Ya

The differences between individual and group behavior are profound, and underpin all the main political and social aspects of civilization. No study of the evolution or operation of our society can hope to explain anything without taking account of how our behavior changes when we assemble in groups. By group I don't mean the instinctive alliances of individuals of a particular species that we call clans or tribes or societies. I'm referring to associations of people within the overall group of society that form spontaneously or by the coercion of an individual. Some groups are long-standing and become highly organized, the best-known examples of which are religion, the military and political parties.

The observation that when individuals form groups they begin to behave differently has been made by many since ancient times, but was not formally analyzed until LeBon's comprehensive 1895 work. Others, including Freud further analyzed the main features that characterize group behavior. The defining characteristics are easily observed, because they don't involve anything not seen in individual behavior. It's the change in the individual's behavior that defines overall group behavior. What has always attracted attention is the irrational behavior of mobs that form almost spontaneously in reaction to some passionately held opinion or resented action of authority. Such groups, though they are at one end of the extremes of group behavior, well illustrate what's going on even at the other extreme where behavior has become disciplined.

> The opinions that are held with passion are always those for which no good grounds exist; indeed the passion is the measure of the holder's lack of rational conviction. Opinions in politics and religion are almost always held passionately.
>
> *Bertrand Russell*

The rough end of group behavior starts with panic, after which come riots, brawls, a mob chasing a thief and the like. We are all familiar with the alarming characteristics of such crude groups, specially the abandonment of moral and rational considerations ('the turbulence of the mob is always close to insanity'). There are three defining elements: anonymity; contagion; and suggestibility.

An individual joining a mob becomes anonymous by shedding his sense of responsibility for his thoughts and actions. He feels that the group has taken the burden of responsibility from him so that it doesn't matter what he does. If the others do it so can he, even though he wouldn't do it on his own. This sensation is part of the contagious nature of a mob. As soon as one person follows the leader in an action, others join in. Once looting, for example, begins, it rapidly spreads and increases. It's an expression of the willingness to succumb to suggestion – a powerful component of group behavior. Put another way, once in a group, an individual becomes credulous and easily influenced by those leading it and those already part of it. Acting on his own account, an individual might be normally intelligent, deal with things objectively and thoughtfully, make up his own mind and heed his conscience. In a group, he abandons those cultural reservations, the more readily the more violent the mob; violence is contagious. One senses that many people seem almost to look forward to an excuse to shed their cultural inhibitions and turn into impulsive, brutal versions of themselves. The police, for example, recognize 'recreational rioting', referring to people who join a riot simply for the fun of it.

The next level of group behavior includes less spontaneous, more organized gatherings such as protest marches, crazes, open-air concerts or sport fans, a controversial socio-political matter such as climate change or abortion, social isms, e.g., pacifism, environmentalism. With the notable exception of terrorism, these groups are not excuses for violence, though as everyone knows they often descend to it. One can be a mental group member; you don't have to be on the street looting and petrol bombing the police. While these groupings may differ widely in their objectives, they nevertheless display the same group idiosyncrasies, some more strongly expressed than others. There is the same decay of moral and rational behavior, the same sense of anonymity, the submission to contagion and suggestibility, the adoption of formulaic actions and symbols such as placards or logos, all of which cement the element of mental unity. Members of such groups, specially when in public, tend to be childish, loud, rude, emotional, impatient, aggressive, intolerant, self-righteous, and indifferent to others' feelings. Protest marchers tend to play drums, sing folk songs and dance.

At the other extreme are long-standing, organized groups such as religions, armies and political parties. Because they've been around for a long time we've become habituated to them and easily forget that they are based on group behavior. The element of mental unity, obedience, is the outstanding feature of these groups; the submission to formulaic actions that develop into the well-known rituals of, e.g., prayer, marching, uniforms, etc. Ritual (or discipline) is probably the most outstanding feature as it completely dominates the individual, less obviously in political parties, though toeing the party line is always expected. Ritualization is enhanced by the use of symbols – crosses and other religious paraphernalia; flags, insignia, badges etc. Party politics is pure group behavior – adherence to guileless ideals of individual and national prosperity, contempt of morality, inane, weaseled slogans (Build Back Better, Compassionate Conservatism, Yes We Can, Blood And Soil, Bread And Roses,), the adoration of leaders as heroes, emotion-driven rather than reasoned thinking, childish abuse of dissent, regression even to such primitive behavior as brawls.

LeBon emphasized the element of mental unity, obedience, which is common to all organized groups, whether newly formed or long established. We possess an ancient, instinctive tendency to obey authority that is really an NSDT statement in that if we didn't there could be no structure to society at any level. The first step towards functional order in a group of people is obedience to someone or thing – that much is obvious. There is a somewhat notorious experiment, conducted before ethics committees were invented, that makes what is meant by obedience crystal clear. A test person was told that the equipment before him administered electric shocks to a test subject sitting in a separate room, but visible to the shock giver. In reality, the equipment was fake and the receiver of the fake shocks an actor. The experimenters told the shock giver to gradually increase the voltage to test the reaction of the receiver. Even though the subject pretended to be in agony as the voltage was increased, the givers (several people were tested) did not hesitate to increase it when told to. Their readiness to obey what would be when making their own considered choices in isolation unacceptable commands startled the experimenters as much as it did those who subsequently read about it. I was only following orders, as has been the rationalization of many familiar to all of us. The Holocaust is the most notorious recent example, but there are countless examples, not only in military circumstances, but in many spheres of life. The Jonesville mass suicide, Kamikaze pilots, terrorist suicide bombers, all illustrate the phenomenon.

The power of suggestion has been remarked upon by many, as it's a striking feature of our behavior. In a sense, it's obvious when you consider how much of our evolution has been to superimpose non-instinctive cultural constraints on our suite of instincts. A child is born with that suite of instincts, but without the cultural constraints that he will possess as an adult; he has to acquire these as he grows up. He could theoretically acquire them by observation, but that would give him a lot of choice, which flies in the face of the purpose of culture, which is to unify its people in terms of their behavior. If culture is to create a reasonably homogenous society it must be adopted in full by all its members, i.e., they must be obliged to take it on board. A great deal of it is transmitted by suggestion to which children are specially susceptible. A child is pre-ordained to believe what it's told by those adults whom it trusts; it's the principal means whereby cultural rules are established. Repetition, ritual and punishment are the devices used to reinforce it, but much of it begins with suggestion. The susceptibility to suggestion wanes with age until in the adult it's largely replaced by conscious judgment. It never decays altogether and any regression to a more infantile state uncovers it so to speak, as it does in the case of crowd behavior.

The breakdown of inhibition and willingness to obey the leader is strongly reminiscent of hypnosis. Most people are familiar with entertainment hypnotists who can hypnotize a group of strangers on stage and then command

> Credulity is the man's weakness and the child's strength.
> *Charles Lamb*

them to perform all sorts of silly things in a strikingly uninhibited manner. The hypnotic trance though commonly referred to as sleep is something quite different with its own physiological characteristics. The trance subject is conscious and aware of what he has been told to do. The most conspicuous feature of the trance is the suppression of inhibition. The subject loses self-criticism of what he is doing and becomes totally naïve – in many ways he regresses to a childish state. I for one find it difficult not to suppose that charismatic leaders unconsciously use hypnosis as part of their ability to persuade their group members to do what they want.

LeBon's third law – hero worship –is not confined to group behavior though it's a conspicuous feature of it. Hero worship is part of the fundamental susceptibility to suggestion and obedience that underpins the structure of society. Heroes epitomize our ideals. We imbue them with all those attributes that ideally we want to possess (but which in reality no one could ever possess). Thus, groups commonly appoint a hero or a martyr or a figurehead – someone who represents the essence of the group's purpose.

****

Summing up, the changes an individual undergoes when becoming part of a group – including the unorganized baboon category of riots and mobs - have an obvious overall character; a regression to a more infantile or primitive mentality. Those things that we cultivate in our individual selves to become civilized are inimical to group function. They must be held in abeyance while the individual reverts to his childish, instinct-driven basic self waiting to be told what to do.

Group behavior wouldn't exist if the genes for it were not subject like any other behavior to positive selection. To me, the reason for it is obvious enough to rank as an NSDT statement. For groups to succeed they must suppress normal, individual behavior. There can be only one point of view and everybody must share and promote it. Were individual behavior preserved there would be hesitation, dissent, criticism, disobedience, defection, etc., depending on each person's point of view. That sort of conflict would weaken the group's objective. The group demands homogeneity, everybody must join in and endorse the group's interests. Individuality by definition weakens a group. We wouldn't form groups if they didn't work, which explains why we are genetically pre-disposed to suppress our individuality when acting as part of a group. The more unified and therefore homogeneous a group is, the less it queries its purpose, the more likely it is to succeed. It cannot afford to have members demand reasoned or truly objective bases for its activities; everybody must believe, have faith, in what the group is for.

Chapter 3   Neurons and the Brain

Brain n. An apparatus with which we think that we think.
Ambrose Bierce

If the brain was so simple we could understand it, we would be so simple we couldn't
Lyall Watson

It's not necessary (luckily) to go into the myriad detail of the brain here, but the themes of this essay can't be understood without first knowing a few basic things about it, given that all our behavior originates there. The 4000 years between when we started writing our thoughts down and the discovery of what the mind and brain consist of and how they work has produced umpteen screeds of speculation – almost all of which can be dismissed as hogwash. It's been extremely difficult for people to accept that intangible, ephemeral entities such as urges, thoughts and emotions, specially superstitious ones, could have relatively straightforward anatomical explanations. That difficulty has generated a flood of fantasy and foolishness that still infiltrates brain research, much as religion corrupts politics. What we have so far learnt about the brain makes it clear that there's no reason to think we won't eventually be able to match everything mental to brain anatomy and function. Sure, it's immensely complicated and has to be studied at the molecular level; but it's not theoretically out of reach.

The Brain

A convenient way of viewing the brain is to think of it as consisting of three anatomically distinct regions, the sensing, the feeling and the thinking parts; respectively, the brainstem, the limbic system and the cerebrum. The brainstem (in evolutionary terms, the ancient part of the brain) connects the brain to the spinal cord. It relays information between the brain and the body, and controls the heart, breathing and level of consciousness. Processing sensory information takes up more than half the total brain activity, emphasizing how much we depend on what we perceive around (and within) us. For my purpose, it's of paramount importance in that it's the seat of our oldest and most powerful instincts; aggression and reproduction. Part of the

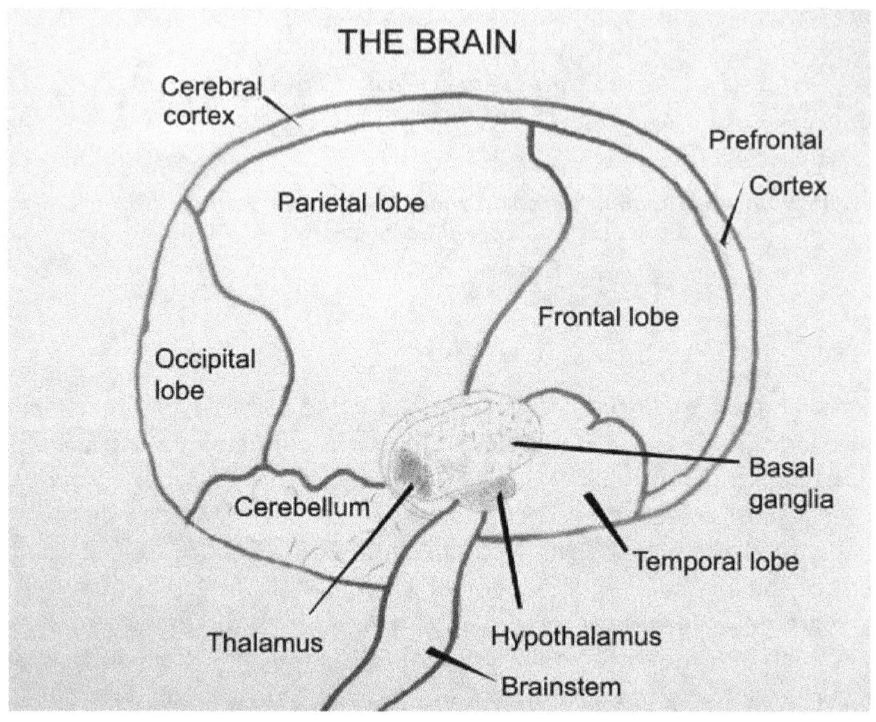

brain stem, the basal ganglia, house the primary instincts such as aggression. The limbic system – the feeling brain - generates our emotions - how we 'feel' in relation to an instinct. In contrast to other animals that respond to an instinct simply by carrying out the associated action without emotion, we have evolved an accompanying emotion that so to speak allows us to 'visualize' the instinct and its consequences. This component of abstraction is a fundamental mechanism of our cultural modulation of instinct that I will return to later on. The emotion itself is largely created by the release of instinct-specific hormones. For example, a sudden hostile encounter triggers the aggressive instinct. Simultaneously, the limbic system generates the corresponding emotion of anger. Hormones such as adrenaline and cortisol are secreted that quicken the pulse and breathing, and increase blood sugar and sweat in readiness for action. The emotion of anger is perceived by the conscious, which decides whether to respond aggressively, or to avoid the threat.

This distinctly human sequence of events involves a group of interconnected structures located deep within the brain above the brainstem. The most important is the *amygdala*, which stimulates the *hypothalamus*, which in turn stimulates the *endocrine* (hormone) system via the *pituitary* gland to secrete the relevant hormones. The rest of the process involves the *cerebrum* (see below). The *hippocampus* encodes new memories.

The limbic system includes several other structures, notably the *thalamus* (see brain waves below), *cerebellum* and *limbic cortex*. The cerebellum, the second largest part of the brain, is associated with the coordination of movements and balance, but also with spatial processing, working memory, language, social cognition, and affective processing. The *limbic cortex* is involved in mood, motivation, and judgment.

The cerebrum, the thinking brain, is by far the largest part of the brain. It's divided into two hemispheres, each of which has four lobes: *frontal; parietal; temporal;* and *occipital*. The rippled surface of the cerebrum is called the *cortex* (from the Latin for bark). The cerebrum and its cortex have expanded during our evolution far more than the rest of the brain – a reflection of the relative importance of 'cleverness' in our evolution.

The frontal lobe is where abstract functions such as reason, thought and planning take place, decision-making, consciousness, speech, memory and emotion and the control of voluntary movement.

The temporal lobe is involved in memory, language, music and emotion. and manages some visual and auditory information.

The parietal lobe integrates input from the senses and is important for perception, spatial orientation and navigation.

The occipital lobe, at the back of the skull, is responsible for visual processing.

The cerebral cortex, which covers the cerebral lobes, is the most recently evolved brain structure, particularly the *prefrontal cortex* where the so-called higher intellectual functions, such as thinking, reasoning, awareness, will and language, are controlled. In fact, what makes us different to other apes is concentrated in the cerebral cortex. The synaptic connections in the front part of the cerebral cortex enable conscious decisions about whether or not to react to a stimulus with action. Specific parts of the prefrontal cortex control inhibition, while others control arousal and movement.

While there's considerable functional overlap among the many brain structures, and the overall coordination of function is extremely complex, basic functional divisions are discernable. The first responder to an environmental stimulus is our armory of instincts located in the brainstem. The limbic system simultaneously adds the appropriate emotion to the instinctive response, i.e., instincts and emotions are separately aroused and controlled.

An unusual structure in our brain is a direct connection between the brainstem and prefrontal cortex that allows us to immediately weigh up (both unconsciously and consciously) the consequences of instincts that have been aroused and either endorse, inhibit, redirect or suppress them. This connection bypasses the limbic system so that regardless of whether or not we have suppressed an instinct the associated emotion remains. An example will make this clearer.

Suppose that you are walking home having just bought a bottle of wine. You round a corner and see a man trying to steal the handbag of an old woman in a wheelchair. You instinctively react aggressively and run to attack the thief, who punches the woman and makes off with her handbag. This enrages you still more. You quickly catch up and have the urge to hit him on the head with the wine bottle. Something inhibits this surge of violence and instead you tackle the thief to the ground, where another person helps you overpower him. You find yourself still as angry as you were initially, even though your initial instinct for violence has subsided. It's some time before your anger dissipates (i.e., the hormones involved disperse).

The same basic process manages the instincts and emotions of the copulatory function and the associated emotions we perceive as lust, love, affection, friendship and so on. Many of these feelings are not generated solely by endocrine hormones but include such hormone-like substances as the neurotransmitter dopamine, which increases when we anticipate something pleasurable. The brain also produces the so-called love hormone, oxytocin, and others such as vasopressin, when you hug someone, have an orgasm or are involved in any sort of social exchange. The brain also releases endorphins that reduce pain and have a sedative or euphoric effect in the appropriate circumstances.

As we all know, we are sensitive to a huge range of intensity of emotions. We rarely perceive only the basic instinct and its emotion. More often, we sense all sorts of levels of feeling that we classify into a huge repertoire of related emotions. For example, arousal of the aggressive instinct will be accompanied by an emotion that will be defined by the context of the instinct.

Were aggression sparked by an attack the accompanying emotion could be strong fear or anger, or even both. Were it sparked by a confrontation with someone where attack was not at issue the associated emotion could range from rage, anger, annoyance, irritation to impatience.

There is also a social decision-making part of the brain where the parietal and temporal lobes meet. This is where group-related actions are managed; actions that directly affect other group members, but where the benefit to the maker is typically indirect and postponed (see Chapter 4).

The Neuron

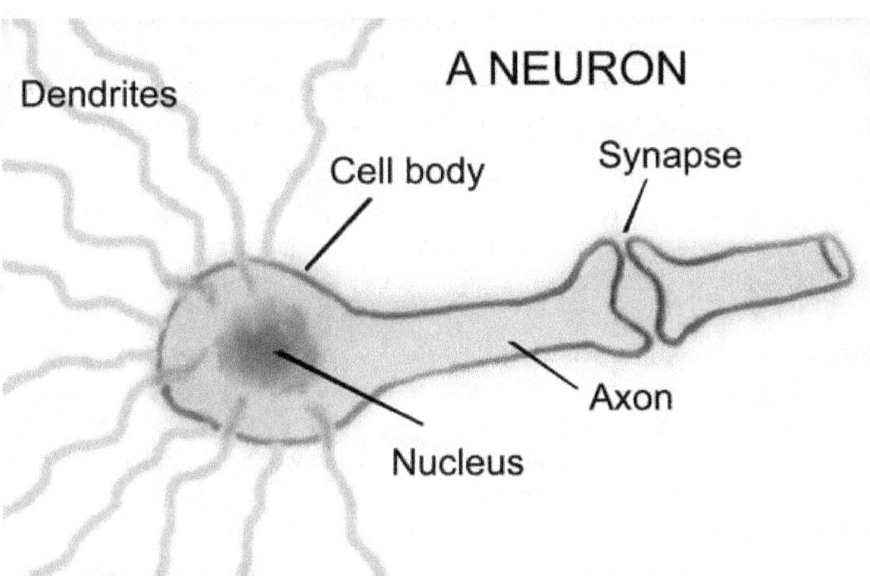

Many other specialized parts of the brain have been identified, but I won't go into those. What matters here is that everything the brain does is mediated by one kind of cell – neurons - of which we have around 86 billion. Every stimulus from a sense organ and every signal from the brain to do something is transmitted by neurons. The bulk of the brain is made up of glial cells that support the neurons physically and nutritionally (though they may also take some part in neuronal function). There are no other mental cells in our brain, though different parts have radically different kinds of neurons (10,000 different kinds in fact).

A neuron has a **cell body** with a nucleus and clusters of **dendrites** that connect with many (up to 15,000) other neurons, creating neural networks totaling 100,000 km. The cell body also has an elongated **axon** that generates a **resting potential** of -70mV across its surface. A message is transmitted by the axon taking up positive ions that cause it to discharge as an **action potential** of +40mV. This current can only travel one way, which means that sensory and motor nerves are arranged in opposite senses: motor impulses from the brain to an organ; sensory impulses from a sense organ to the brain.

Every axon is interrupted by one or more **synapses** – a narrow gap - before it reaches its destination. When an axonal current reaches a synapse, it has to somehow 'jump' that gap to continue. This synaptic gap is the principal device that gives the brain its capacity to do so many different things. Without it, an axon once 'fired' would send its message irreversibly and unchangeably, i.e., in a simple, predetermined way. The synaptic gap evolved as a switching complex that allows the axon's current to be modified in many ways (bearing in mind that there are 100 trillion synapses 'many ways' is an understatement).

When a neuron's current reaches a synapse it stimulates the secretion of a chemical (called a **neurotransmitter**) that crosses the gap and 'fires' the continuation of the neuron. But, that is only the simple situation of a nerve impulse travelling along a single neuron. Many subtle things happen at synapses, particularly those in the brain, that weaken or strengthen the intensity of the axonal discharge, or delay or modify it, or stop it altogether. The neurotransmitter may cross to stimulate the axon of another neuron that connects via dendrites with other synapses. Not only that, but nerve impulses employ bunches of adjacent neurons for their transmission, rather than single neurons. Thus, many synapses may be involved in a given impulse, further increasing the possible outcomes, as not all synapses act in the same way. Moreover, more than 200 neurotransmitters have been identified, each of which has a specific function. Quite obviously, the capacity to modify a nerve impulse is potentially astronomical.

Though there are many different sorts of synapse they are all either excitatory or inhibitory, and every neuron has both types located in different zones. There is an excitatory dendritic cluster and an inhibitory cell body. Typically, excitation has to overcome inhibition before it can reach its final destination. This can become subtle and complicated in the case of repetitive actions. For example, acrobats' brains learn what movements to expect, and inhibitory neurons cancel out reflexes that would ordinarily prevent one from performing such bizarre movements.

Much of pathological behavior is the breakdown or absence of inhibition so that wanting and wishing manifest at full-strength. *I see, I want, I take*. Since the wanting is instinctive and therefore invariable (in all of us), whereas the inhibition is largely cultural, it's not surprising that inhibition may be lacking, or decay. If it was never properly installed in its cultural template(s) it simply doesn't operate. And being stored in a memory-like process, it can decay if not reinforced. Statistically, by far the commonest failure of the cultural brake on *'I see, I want, I take'* is theft, followed by assault.

This is where Freud's most important insights are beginning to be explicable in terms of brain function. His interpretation of our behavior as consisting of excitation modified by inhibition well matches the way in which neurons function. The actual neurons involved are already being identified so that there's no reason to suppose that one day we will know the complete neural networks of instinct and their culturally induced control, and of thoughts, emotions, beliefs and so on.

In addition to the action of neurons there's one other important brain function: **brain waves**. These are rhythmic, oscillating patterns of electrical activity across the whole or large parts of the brain, characterized by their frequency, amplitude and phase. Many are generated by a thalamocortical loop, a two-way circuit that connects the thalamus with parts of the visual cortex. Others are generated by neurons as rhythmic patterns of action potentials or spikes. Spiking patterns are considered fundamental for information coding in the brain and brain waves generally are considered to be the mechanism of transmission of perceptual groupings such as the shape and color of an object, the sense of time and other general as opposed to specific kinds of information. A lot remains to be discovered about brain waves.

Memory. I will briefly outline memory, which is important because everything we do is influenced by what we have memorized since we were born. Everything. Every instinctive urge is modulated by comparing its consequence with what we already know – in the widest sense. What we know (or think we know) about ourselves, our kin, our society, the world. We do not act in randomly novel ways – at least 'normal' people don't. The seemingly random or erratic behavior of the mentally disturbed emphasizes that the intact mind produces overwhelmingly predictable, explicable behavior. Not necessarily appropriate behavior from others' points of view, but strongly felt to be 'right' by the thinker. And not necessarily predictable from an observer's point of view because that is strongly influenced by what the observer wants to see.

A memory is made by 'locking' the chemical changes at the synapses that transmitted the perception remembered. As everyone knows, the event may be remembered for only a short time – the short-term or working memory that can hold around seven perceptions at a time for a few seconds. Depending on their importance short-term memories may quickly be forgotten, or be added to the store of long-term memories that can last a lifetime. Generally, memories are consolidated by repetition of a perception, as in intentional (or unintentional) learning, that repeatedly stimulates the same or similar perception. Repeated stimulation of a particular pattern of synapses strengthens the synaptic signal, both by increasing the amount of neurotransmitters and by increasing the number of receptors on the post-synaptic neuron.

We have a huge capacity for remembering –something of the order of 100 trillion connections between synapses are associated with memory creation. Synapses increase or decrease in strength and size when activated, creating in effect a coding analogous to the 1s and 0s of binary computer coding. Not only do we store far more memories than we can consciously recall but the brain is constantly revising, consolidating and condensing what we've memorized each day, as well as older memories. Many memories if not repeated or recalled decay and are forgotten or distorted. Some memories can be directly recalled by the conscious, but the bulk are only accessible by association. You might want to recall someone's name after a long interval since you met him. By associating the meeting with other events relating to the time and place you can eventually arrive at the long-forgotten name. This process of proceeding along a chain of associated events to arrive at a deeply imbedded, hard-to-recall memory is in essence the procedure that Freud used to develop psychoanalysis.

One of the many functions of sleep is to filter the day's memories and condense them. What is left after condensation strips away the fluff is perceived more clearly and is the main step towards the clarification of complex memories and the formation of concepts. This process explains why things often become much clearer the morning after, or days after, while trying to sort out a complicated matter. It's why one shouldn't revise just before exams. The cooling off period in law is one of the all-too-few things about memory that the law has incorporated.

Though we are apt to treat our memories as an exact record of what we have experienced, the truth is that they are not at all an exact record. Freud noted that memories are subject to strong biases depending on the nature of the event. There are two general categories of bias: that caused by wish fulfillment; and that caused by traumatic events. Freud referred to 'The great Darwin' who said '*I had during many years, followed a golden rule, namely, that whenever a published fact, a new observation or thought came across me, which was opposed to my general results, to make a memorandum of it without fail and at once; for I had found by experience that such facts and thoughts were far more apt to escape from the memory than favorable ones.*' Nietzsche's aphorism neatly sums this up: 'I did this,' says my memory. 'I cannot have done this,' says my pride and remains inexorable. In the end – memory yields. Freud considered that testimony in court does not pay enough attention to the fallibility of memory, and, gradually, judges are taking note of mnemonic flaws and modifying juridical procedure accordingly. It's now well known that eyewitness memories of, e.g., accidents, are rarely photographic records, but often quite different to what actually happened. Furthermore, they can unconsciously change in response to what others said they saw, or what the viewer would like to have seen.

Traumatic events, and things that are generally perceived as negative, attract more attention from the brain, which typically detects negative information faster than positive information. This manifests in its extreme form as shell shock (which the *jargonista* spin as post-traumatic stress disorder). It would be an NSDT statement to say that selection favored this sort of bias as a protective device to help avoid harmful circumstances.

****

Summing up thus far I realize that up to now I've said little about the purpose of this enquiry. That's because I think it's impossible to explain the purpose without knowing something about the way in which we think and why it evolved the way it did. I've tried to describe our behavior as the outcome of long-established instincts residing in the unconscious part of our mind, whose expression is edited so to speak by our conscious minds to result in outcomes that are better suited to life in a social group. I've reviewed some of the many mental devices the conscious uses to conceal the self-serving nature of what we do and make it appear that they serve wider social outcomes. I've briefly discussed the brain's electro-chemical apparatus for storing memories and for inhibiting raw instinct. I want now to go over our evolution to show how over many millions of years we have gradually evolved ever more effective ways of doing things cleverly as opposed to forcefully, and how this has been achieved by extensive adaptation of our conscious and its mediation of our instincts.

For those unfamiliar with the genetic weapons of this contest with our environment I will make a brief digression into the genetics of evolution. It's not optional knowledge, because we have modified the way in which our genes work to a degree and in a manner not found in other animals, not even close relatives like cousins chimp and bonobo. Without appreciating these modifications and their relation to fundamental genetics, there's no chance of understanding our behavior. But, first, the basics.

Chapter 4   Genes and Natural Selection

Every one of the 100trillion cells of our bodies including those that harbor our instincts, sensations, memories, and thoughts carry out their assigned tasks based on instructions from the relevant genes in their nuclei. So easy to say, so much going on. This is the molecular level of life; a world so different from the scale of things that our eyes evolved to see as to be almost unimaginable. I certainly find it hard to visualize things the size of molecules and indeed, we are only now getting some idea of what this world looks like. We are visual animals and want so much to *see* everything, even though those who study genetics have trained themselves to figure out what's going on without visualizing it conventionally. All the same, I find it stimulating to know that everything, but everything about our bodies, behavior and evolution is determined by just a handful of molecules so tiny that they all fit comfortably inside a cell nucleus. Just how these bits of DNA, our genes, command the huge variety of life on earth is to me a fascinating story, with which I hope the following brief outline will make you agree.

Genes are precise segments along a long molecule of DNA. This DNA molecule is supported by a ladder-like frame of other molecules, making a long, tightly coiled strand called a **chromatid**. This is the famous double helix of DNA that most of us have heard about. We humans have 23 unique chromatids supporting around 22,500 genes (the exact number is not yet known), that constitutes the human *genome*. These 23 chromatids don't exist singly; rather they are always in identical pairs joined at a particular point (the **chromomere**). These pairs are our **chromosome**s. Every human cell nucleus has the same 46 chromatids of DNA assorted into 23 chromosomes. These are tightly coiled to pack them into the cell nucleus (if they were straightened and laid end-to-end they'd extend 6 feet). Each time a cell divides to make a new cell it makes an exact copy of every chromosome for the new cell.

The reason why chromosomes are made of two identical chromatids becomes obvious at the time of reproduction. When the reproductive cells – eggs and sperm – are made, the chromosomes are pulled apart so that each sperm or egg has only one chromatid of the respective parents' chromosomes. At fertilization - conception - the single male and female chromatids unite to restore the normal 23 chromosome pairs. Because the parental genes are dissimilar, the offspring chromosomes aren't identical to either parent's. That's why offspring resemble their parents, but are not identical to either. Most of our features are determined by the combined effects of several genes rather than just one. During the splitting and recombining of chromosomes to make a fertilized egg, various things happen that result in alterations to the final gene complements. These alterations add to the differences between parents and offspring and among siblings.

A little more about DNA and genes may be helpful. A molecule (chromatid) of DNA consists of a long chain of pairs of just four **amino acid bases**: thymine; adenine; cytosine; and guanine, conventionally depicted as T, A, C, G. These bond in only two possible pairs; either a T with an A, or a C with a G. Each pair is called a **base pair**. Each base pair is bonded to a sugar-phosphate frame molecule to form a **nucleotide**. Each nucleotide is attached to another to form the helically coiled strand of DNA, the chromatid, described above. This simple enough arrangement starts to get hard to imagine when you discover that a single molecule of our DNA, a single chromatid, averages more than **half a billion** nucleotides. What's surprising (at least to me) is that only 3% or so of a chromosome's DNA comprises genes. That's all - 3%. About 80% of the rest acts to switch specific genes on or off, or vary the intensity of their effects. The 17% residue is regarded as inert DNA from old genes and viruses (though it may yet turn out to be functional).

A gene is a particular segment of DNA along a chromatid, a specific number of adjacent nucleotides (averaging about 15,000). The same unique chunk will be found in the same place on every copy of the relevant chromosome. In us, each chromatid supports an average of ±1000 genes. All 7.3 billion of us carry the same 23 chromosomes, containing on average 99.9% of the complete set of human genes – the human genome. At the same time, every chromatid is unique (every chromatid of every organism, not just us) even though they all are made up exclusively of the same two base-pair units - around 650 million of them. Uniqueness is not owing to the 0.1% (around 20) absence of a few genes – many of us lack the same few. What confers uniqueness is the **order** of base pairs along a chromatid. It's easy to see that the possible differences in the order of over half a billion base pairs of DNA are astronomical. Moreover, the chances of any two chromatids being the same are remotely small. Put another way, the 0.1 difference between us translates to an average difference of 3 million base pairs; things at this scale can be tricky to visualize!

There's a slight complication here (in genetics there are **always** complications!). Most genes exist in two or more slightly different versions, called **alleles.** All alleles will be located where the gene they represent is located on a chromosome - but only one allele at a time. We may inherit the same allele from both parents, or two different alleles. The commonest are **dominant** and **recessive** alleles, meaning that for an individual to possess a certain characteristic it must inherit a certain combination of the dominant and recessive alleles. Eye color and blood types are both examples of various dominant and recessive allele combinations. Alleles that become so common that they affect more than 1% of a population are called **polymorphisms** (for example, the human blood types A, B, AB, and O). Most polymorphisms have little or no effect on the **phenotype** (the actual structure and function – the outward appearance - of a person's body).

DNA is tough stuff and, left to themselves, genes like diamonds are forever. Occasionally, when a bone or tooth fossilizes in just the right cold, dry conditions archaeologists have been able to recover almost intact pieces of chromosomes – even some complete chromosomes. The oldest recovery to date is from a 1.2 million-year-old mammoth, but the oldest human DNA recovered is only 45,000 years old. Tough though DNA is, copying mistakes, viruses[6], injuries, radiation and some chemicals can alter a gene's composition so that it **mutates** into a new, different allele. Mutations are also caused by the insertion or deletion of DNA segments, transposons, which can move around the genome. Mutations that occur in your body cells during your lifetime may cause a disease such as cancer, but cannot be passed on to your children. But, if a mutation occurs in a sperm or egg chromosome it will be passed on, and may affect how the individual inheriting that allele deals with life. Most mutations are harmful - many fatally so – while others have no noticeable effect. Every now and then (about 1 in 10,000 mutations or one mutation in every 30 million base pairs) a mutation will make the inheritor better at something – abstract thinking, digesting certain foods, knapping stones - anything. Darwin saw it as making the individual fit better into its environment, hence his term **survival of the fittest.** He and Wallace correctly concluded that the natural selection of those changes that better **adapted** the individual to out-compete its fellows **(survival of the fittest by natural selection**) was the mechanism of evolution, even though no one yet knew of the existence of genes. It was a smart bit of thinking.

What Darwin and Wallace said in 1858 can be paraphrased as follows. A beneficial mutation will gradually be inherited by all descendants of the original holder, until eventually all members of the population carry it. It's then said to be **fixed** in that population and will stay fixed as long as that population

---

[6] There is evidence that around half our genome resulted from mutations caused by viruses.

survives. Fixed genes cannot be selected against, as natural selection is a one-way process. The only way a fixed allele can be lost is by genetic **drift.** Gene fixation is why we all have some genes that first mutated into existence billions of years ago as well as those we came by as we evolved. Around 60% of our genes are the same as those of, for example, a banana; we share 80% with mice, and up to 99% with cousin chimp. The really old genes we share with distantly related organisms are for such things as basic proteins required for universal functions such as cell structure and energy turnover and other fundamental functions. They persist because mutations to them are always fatal.

There are two blips in the above description of natural selection that affect the overall outcome in animals such as us. The first concerns the role of males in sexual reproduction. For a mathematical reason independent of natural selection sexually reproducing vertebrate populations will always approximate equal numbers of males and females; yet for reproduction most of those males are superfluous since one male can (and often does) fertilize many females. This has resulted in a selective advantage in being good at siring successful offspring that perpetuate your genes. Since many males will possess such attributes selection will also have favored those males that compete most successfully with one another for first access to females in estrus (wildebeest and elephant seals among many). Sometimes the competition is not directly between males, but instead for the 'favor' of a female who only accepts her perception of the 'best' male (e.g. courtship displays of birds of paradise among many). Natural selection thus shapes a species' overall population and simultaneously the sub-population of males. This parallel **sexual selection** as it's called is not entirely independent though, because 'dominant' males must still compete with all members of the population for food and overall survival attributes.

The second blip has had a profound effect on us humans. It's that some mutations caused individuals to behave in ways that benefitted the whole group of which they were members rather than just themselves. Such mutations caused what is somewhat misleadingly called altruistic[7] behavior such as sharing food, or helping another to raise offspring, or performing joint activities such as fighting. These are prominent features of our social behavior and have caused an immense amount of controversy, because they seemed to contradict the basic premise of evolution: unless a mutation confers an

---

[7] Altruistic really means being intentionally selfless. In reality, most inclusive fitness attributes are instinctive. It's possible that we are capable of true altruism; but more likely that most of our apparently selfless behaviour is in fact instinctive. Such matters remain to be resolved.

advantage to the possessor it's simply impossible for it to spread because to do so that individual must be more successful at breeding than his fellows. If he voluntarily helps his fellows to fight off a predator before he has reproduced, and is injured such that he cannot reproduce, how can his altruistic gene have benefitted him? The answer is that his altruistic behavior improved the survival of his relatives' progeny as well as the whole group, which in turn meant the genes he had in common with them had a better chance of perpetuation. Demonstrating that this **inclusive fitness** form of selection actually happens has proved extremely difficult, but it's now generally accepted that inclusive fitness is a special case of natural selection in social animals that does not contradict the basic premise of natural selection. The individual's fitness is indirectly improved by the mutation for altruistic behavior.

That's a bare-bones description of the mechanism of natural selection. It happens because genes are constantly mutating (the DNA passed to each new generation carries an average of 150 new mutations) so that the human *gene pool* as it's called is constantly changing. This natural variation means that some individuals will, by chance, fare better in comparison to others. As changes accumulate the anatomy, function and behavior of the population of individuals may collectively change so much that a new species evolves. In a static environment the net changes would be negligible over time. But, the environment is never static and the pressure of natural selection is proportional to the degree of environmental change each individual encounters.

It's vital to remember that in all this Nature doesn't give a damn what happens to genes or to the individuals carrying them. The forces that change genes act at random. Whether an individual lives or dies as a result is also a matter of chance. There is nothing controlling the consequences of mutation. This makes it difficult to talk about natural selection in a truly detached, objective manner without being foolishly pedantic. The term natural 'selection' already implies some sort of choosing. We say that Nature 'selects' those genes that make an individual fit into its environment 'better' when really Nature does no such thing; it's just a bystander. Nothing in Nature is better or worse; nothing is being chosen or ignored. We came into existence by chance and will like all species vanish the same way (unless of course we collectively commit *sati* on the funeral pyre of civilization). But, to describe evolution by natural selection without recourse to anthropomorphic rhetoric is well-nigh impossible; fortunately, it doesn't matter as long as we realize what we're doing, and that natural selection has no ordained direction, no strategies, no inevitable outcomes. We are not some sort of immaculate confection, predestined by fate (or god) to occupy the 'pinnacle' of evolution. To suppose that we are the 'highest' life form, better than the rest of evolution - more highly evolved - is a purely human, narcissistic notion. Nature has no such ranking and is unaware of our lamentable self-admiration.

That's the process of evolution at its simplest. All sorts of subtle and complicated things are going on that modify the basic mechanism, not all of which are fully understood by any means. For the present purpose it's enough to know that this is essentially the way in which natural selection of 'new' genes (alleles) works. A beneficial gene is passed on to the possessor's offspring, and then to theirs, so that gradually it spreads through the population until all members have it. (You can skip the remaining paragraphs if you've had enough of genes and natural selection.) Mutation is the only process that changes a gene, but the distribution of genes in a genome is affected by two categories of process: molecular events that affect genes directly; and those that affect genomes indirectly. The first includes several quite different but universal processes.

I've deliberately not described how genes execute their instructions for making cells develop in the myriad ways necessary to construct an operational organism. That's a hugely complicated chain of molecular events, and not by any means fully understood. Luckily, it doesn't affect my purpose here, save for one aspect of it. That is the switching on or off, or varying the effect of a gene. The bulk of a chromosome's DNA appears to be for this process that has far-reaching consequences. Not only can it result in individual variations in how a gene works, but it also means that the same allele in different species or populations can have different effects. Thus, while we and cousin chimp share 99% of our genes, it doesn't follow that a given shared gene will have the identical outcome in both of us. It might be switched off in one of us, or have a different effect because it's associated with a different suite of genes. A lot depends on what other genes each of us has and in what ways their actions are controlled at the gene level.

When reproductive cells are made and the male and female chromatids join up, some alleles **cross over** and **recombine** with the opposite chromatid, effectively shuffling an offspring's deck of alleles. This is one of the main ways in which genetic variation occurs. There are others, including genes that 'hitchhike' on neighboring genes, so-called 'jumping' genes, and others still. As I said, there are always complications in genetics that blur the basic processes.

There is an important occasional process that acts to sharply reduce the genetic variation in a population and that is when circumstances combine to kill off the majority of its members. These catastrophic events are called crashes by zoologists and **bottlenecks** by others. They result in much lower overall variation and a different suite of allele frequencies, as many alleles will be lost altogether. There is evidence that we underwent at least one and possibly several bottlenecks during our evolution that might account for some of our most prominent features.

The **founder effect** occurs when a relatively small portion of a population separates from the main population to found a new population. This new, small population will inevitably have less genetic variation than the larger, parent population, as it is in effect a sample of the main population.

The second category, processes that affect whole genomes, comprises **drift** and **gene flow**. Drift is the inevitable, random loss of unfixed alleles that happens if for some reason a mating fails to produce any offspring. It can result in the fixation or complete loss of an allele independently of natural selection. Its effect is slight in large populations but will be quite strong in small populations. Once an allele becomes fixed, genetic drift for that allele stops.

Gene flow is simply the introduction or loss of genes as a result of individuals entering or leaving a particular gene pool. Within a species, separate, or partially separate populations will evolve independently and gradually acquire slightly different genomes. These are somewhat arbitrarily labeled populations, varieties, races, subspecies, species and so on. The genomes of species are ideally so different from their parent species, or common ancestors, that they cannot interbreed. In reality, mammal species that radiated from a common ancestor can interbreed for 2 My or more, even though they may look and behave quite differently. For example, we and the ancestor of Neanderthals split to evolve independently around 500,000 Ya and are regarded as distinct species. Yet there's clear evidence that we interbred with Neanderthals, because we carry up to 4% of their genes (as they correspondingly did of ours).

****

To sum up, there are three reasons why I've given this brief outline of our genetics. Firstly, I wanted to emphasize that everything about us is fundamentally dictated by our DNA, which instructs our cells, all of them including our mind-cells, to function in a set, predictable way (not that we've figured out all those processes; far from it. But, given the time we could). Secondly, I wanted to emphasize that we and all living organisms have evolved into the forms we see today by a long, long accumulation of mutations to our genes. The rate at which such mutations were incorporated into our genome is such that it took 19 million years from the time our ancestors evolved into the clearly traceable hominid clades to reach our current form. Nineteen million years. Three quarters of a billion generations. Definitely not a one-day job. We speak of history as going back thousands of years; but that's not history — that's just yesterday's papers (a small clip from the evening edition what's more).Lastly, I wanted to say that although quite recently we evolved the ability to modify our instinctive behavior via cultural controls, that ability itself could only have arisen by natural selection of the enabling gene mutations. Nothing about our behavior can be understood without appreciating that. Our genes are our Great Helmsmen; it pays to know their habits.

Part 2 Our early evolution

24.5 Mya – 2.6 Mya

Chapter 5   The Miocene and Pliocene

One of 100 Miocene apes gave rise to our clade seven to eight Mya – we and other Great Apes then parted company – temperature initially much hotter but declined throughout Miocene and Pliocene – fist of our kind of ape, the bipedal Australopiths, evolved late Pliocene – little known of their lives as they left no artefacts – one of them evolved into our ancestor, Homo habilis.

To fully understand our behavior it's important to appreciate that we are the product of a long, long evolution that has accreted a genome containing many ancient genes. Genes don't die or wither away or change their spots. Once fixed in a species, they're indestructible[8]. Like Old Man River, they just keep replicating along. These old-timers are the majority and will always assert their influence regardless of what any Johnny-come-lately genes might do. And, yes, we do have some new-fangled genes, the ones that led to our brain enlarging that led to the cultural ruses for weaseling round the bullying instinctual old-timers. But, unless one sees how little evolutionary time has gone into the newcomers' genes compared to the greybeards' one cannot properly appreciate the latter's authority.

Our origins are also intrinsically fascinating, because we have an abiding compulsion to know why and how we came by this behavior or that. And all we have to go by is the erratic, fragmentary fossil record – little bits of enticing information that we try to piece together to make a cohesive whole. Each new discovery is seized upon by the experts and advanced as evidence for one hypothesis or the falsity of another. The battle of the experts is a hotly contested one and probably unwinnable because there will always be more fossils to unearth and unexpected features to discover.

Although mutations occur randomly, selection is determined by the circumstances of the day. If those circumstances never changed, evolution would progressively slow down as fewer and fewer mutations conferred any benefit. But, the circumstances - the environments - are never static. The

---

[8] A fixed gene can occasionally be lost due to genetic drift.

environment is everything that surrounds us, the physical components and the other organisms and their behavior, as well as everything within us. All can influence the selection pressure on a new mutation, i.e., how rapidly it spreads through the population. In general, the strongest selection pressure is believed to be exerted by climate. During our evolution the overall average temperature has declined steadily from around 19°C 19 Mya to the present 14°C. Within this overall trend, the climate has repeatedly oscillated from hot and wet to cold and dry. Each time it does this, the conditions favoring this or that plant or animal vary enormously.

Our origins of course go right back to the first living molecules, as do all life forms. But, I'll start with the ancestral forms of the Great Apes, the hominids, since they are our grandparents (many times removed). They evolved 19 Mya during the Miocene Epoch. It took another 11 million years of evolution before the common ancestor (whose identity we can't be sure of) of us and our nearest living relative, chimps, appeared, still walking on four legs. Three million years later some apes had started walking about on their hind legs, at which time (6-8 Mya) we parted company from chimps and the other Great Apes – bonobos, gorillas and orangutans – and headed off on our two back legs down the evolutionary track towards the weirdest ape of them all. It took another 3-4 million years for the first weirdos, or human-like apes –the australopithecines - to evolve and seal our fate forever.

The Miocene

At least 100 species of ape lived during the Miocene, ranging over Africa, Asia and Europe and varying widely in size, anatomy, diet and no doubt behavior. We don't know which ape or apes founded our clade (i.e., the animals we share a common ancestor with), but it or they lived between 7 and 8 Mya. The first bipedal apes of our clade appeared in Africa at the end of the Miocene, including such tongue-twistery types as *Sahelanthropus, Orrorin*, and *Ardipithecus kadabba*. This was also the point at which cousin chimp, unimpressed by the new-fangled craze for swaggering about on two legs, waved goodbye and continued on all fours in the forests, the remnants of which they still live in.

The average global temperature declined during the Miocene, except for a three-million-year-long period 18-16 Mya when the temperature abruptly rose to around 18°C (four degrees hotter than it is today). Thereafter it continued falling again. This resulted in a drier climate that saw a trend to more open savanna, characterized by gritty, fibrous, fire-tolerant grasses. It led to the evolution of grazers such as horses, hippos and long-legged gregarious ruminants such as antelopes and the concomitant extinction of many browsers. Ninety-five percent of modern plants had evolved by the end of the Miocene. A large asteroid hit the earth sometime in the epoch leaving the 52 km-wide Karakul crater that may have coincided with and contributed to the decline in temperature.

Timeline 1 summarizes the main events of our evolution as they concern us here, and when they occurred. The 'first' occurrence of an event must always be taken as provisional. Not only are these dates frequently revised to reflect new discoveries, but the assumption that what has been unearthed is indeed a 'first' occurrence is self-evidently improbable. It's simply the earliest found and the chances of it being the first of its kind extremely remote. Which is to say that the oldest fossils of anything are really only evidence that that thing had already been around for quite some time.

Timeline 1.   **The Miocene and Pliocene**

Mya	Event
2.6	End of the Pliocene and start of the Pleistocene.
3.3	Start of Lower Palaeolithic (The Old Stone Age). Ends with Middle Stone Age 300,000 Ya
3.4	Animal bones with butchering cuts (Ethiopia)
4.5	First *Australopithecus* apes in Africa. Larger brain. First stone tools (Ethiopia)
6	End of Miocene start of Pliocene. First bipedal apes
8-4	Our ancestors split from chimp ancestors, but interbred for a long time. Another strong temp drop, Antarctic ice nearly as thick as now.
9	Gorilla ancestor splits from man/chimp clade.
14–12	Strong temperature drop- the Middle Miocene Disruption
19	Ancestors of Great Apes: humans, chimps, gorillas, and orangutans—the hominid clade
28-25	First apes (early Miocene), bigger brain than monkeys
40-36	First monkeys (early Oligocene), colour vision, opposable thumb, social behaviour

The Pliocene

The global average temperature at the start of the Pliocene was around 16°C, some two to three degrees warmer than today. Atmospheric CO^2 levels were similar to today's. By the end of the epoch 2.7 My later the temperature had dropped to less than 15°C. The steady cooling and drying continued the trends of the preceding epoch with the shrinkage of tropical forests and the expansion of grasslands and savannas. Deciduous and coniferous forests proliferated and deserts formed. Mid-latitude glaciation began and during the succeeding Pleistocene epoch global cooling continued, to create the most recent of at least five great ice ages that have occurred on earth.

This was the environment in which we and at least 30 other two-legged apes evolved, as did our four-legged cousins – chimps, gorillas and orangutans. The latter wisely kept their heads down, reckoned we could have the hot savanna with its lions and hyenas, and made for the cool, shady forest where they could relax in relative peace. They couldn't have known that once we had trashed the savanna we would return and cut their forests from over them. Evolution doesn't tell anyone their fortunes; Mother Nature doesn't care.

Bipedalism was all the rage and the 30 or so proponents of the new fashion had their respective days for a million or so years. One by one they toppled into the garbage bin of extinction until finally only one remained, the smartest of them all, as our conceit would have it. How we pulled that stunt off is by doing something I think is unique in evolution. It would be nice if the stunt was an admirable one, as, rather pathetically, so many of us think; but those who value perceptions of truth and honesty, beauty and poetry; of wilderness, of Paradise, of the Garden of Eden, regard it as ignorantly selfish, violently destructive, and altogether abhorrent. But, I'm getting ahead of myself and will return to this later on.

The first animals that are considered to be our kind of ape, the australopiths, appeared in the late Pliocene about 4.5 Mya. For better or worse they heralded the evolution of the ape known officially as *Homo sapiens* (unofficially. *H. stupidus*). These little people – they were only 1.2-1.4 m tall, weighing 31-50 kg, or the size of an 8-9 year-old human – had a good innings before they too went extinct some 3 million years later. Though they were widespread in Africa, and left many fossils, we know almost nothing of their lives because they had nothing but their bones to leave behind. That means they lived largely by their wits, much as chimps still do, aided by only the simplest perishable tools. Though bipedal they had relatively long arms and could still climb trees nimbly and probably slept aloft like chimps; they would have made easy fast food for leopards, lions and hyenas[9] had they slept on the ground. They had 20% bigger brains than chimps and some species at least had started using sharp stones to scrape meat off bones. Chimps use stones to crack nuts, but for tasks like eating meat to this day rely on their hands and jaws. Australopiths' jaws and teeth were smaller than their predecessors' – the beginning of a trend in size reduction that led to our almost vestigial capacity for biting and tearing. Some of us can bite off an ear, or a bit of one – but that wouldn't cut much ice out in the bushveld.

It's become a given that bipedalism was the evolutionary event that more than any other paved the way for our evolutionary destiny. An opposable thumb meant that as soon as our hands and arms were no longer tied up with running and climbing they could be used for hitting, carrying and throwing and above all to make things. If our Pliocene forbears made wooden things, they haven't survived so we're in the dark as to that. As chimps make and use around 20 kinds of tools, mostly from vegetation, it's generally assumed that they have done so for a long evolutionary time and that in all probability australopiths made similar tools (though it's possible that it was a recent

---

[9] These predators as we know them today had not yet evolved; many other, now extinct, predators were present.

adaptation in chimps).At some unknown point artificial extensions to our arms such as wooden clubs and spears evolved to become in due course universal human tools. We probably never will know who first used such devices because small bits of wood rarely fossilize. Until we embellished our spears with stone tips allowing us to confidently infer the use of spears we just don't know for how long purely wooden spears had been used. Crude stone blades have been found with australopith fossils, together with telltale cut marks on associated animal bones; the 4.5 million-year-old obsession with knapping stones to make sharp blades had begun.

Part 3

2.6 Mya-600,000 Ya

Chapter 6   The Pleistocene

The famous Stone Age, our very own Grand Ice Age

Our evolution from the first species of the genus *Homo* to the start of the Big Beautiful Brain.

First two million years of The Pleistocene Ice Age, the Stone Age – fluctuating temperatures – rapid evolution, H. habilis to H. erectus to H. heidelbergensis to H. sapiens – the Last Common Ancestor - possible beginnings of abstract thinking – brains get bigger, muscle power declines, reduction in jaw and tooth size – digression on how we coped with predators – hand tools - cooking – loss of hair – black skin - huge increase in sweating for cooling the brain.

In this Chapter, I will only cover the Pleistocene to 600,000 Ya – the point at which our brain began a half-million-year-long phase of enlargement. That event is so important that I prefer to give it special attention (Chapter 7).

The Pleistocene is the geological period starting 2.6 Mya at the end of the Pliocene and ending just 11,700 Ya. Up to now our ancestors had lived in an environment of slowly declining temperatures, a circumstance that was about to radically change; an ice age was beginning. Compared to previous ice ages the Pleistocene Ice Age has (until now, at least) only lasted a short time. Former ice ages lasted tens of millions of years. The Pleistocene Ice Age, also known as the Old Stone Age, wasn't a single climatic event, but a succession of about 40 cold, dry glacials separated by warmer, wetter periods – interglacials. It began about 2.6 Mya and is said, quite arbitrarily, to have ended 11,700 Ya with the end of the last glacial and onset of the current interglacial (officially known as the Holocene; unofficially – at least by me – the Horrorcene). In fact, there's nothing to say that our Ice Age has ended. In the first one and a half million years, a Pleistocene glacial cycle lasted about 41,000 years, while in the last 800,000 years the cycles lengthened to about 100,000 years. There's nothing to say that the present warming, exacerbated by us or not, hasn't still another few millennia to run before the next glacial sets in; it's the logical conclusion. Some say the next glacial is overdue, and others that it will begin in 10,000 years' time. Whether or not our species has that much puff left is likewise unknown. It's certainly not a given, notwithstanding lord god almighty's assurance of life everlasting.

Overall, the Stone Age climate was colder and drier than today's warm (14°C), wet interglacial. Initially the temperature ranged 4°Cbetween the glacial and interglacial peaks at higher latitudes, less towards the equator. When the cycles lengthened the temperature range increased to 9°C in the higher latitudes, with temperatures averaging between 5°C and 7°C . The generally cold Pleistocene climate was a continuation of a long-term trend of continuously falling temperatures over the last 55 million years. During the Pleistocene's coldest periods enormous glaciers formed in Europe, North America and Patagonia. Two things drove their huge extent. Since most of the global water was frozen, rainfall was about half of today's. Winters and summers varied the temperature on an annual cycle, but these were too weak to stop the glaciers advancing, because the summers didn't completely melt the snow so that it formed additional ice each winter. The second was that the continents were so positioned that warm ocean currents flowed north, losing heat and precipitating snow. For extended periods, 30 percent of the land was covered by a mirror of ice and snow that reflected a large proportion of solar radiation back to space, further intensifying the cooling.

The large temperature range of each glacial cycle would have had a huge effect on the fortunes of all the animals, many of which failed to adapt and died out. Those apes that did manage to weather the climate changes must have done so by moving back and forth with the cold front, so to speak. This is reflected in the absence of any physical changes in the fossil remains that would be expected had we and the others stayed put in the coldest parts of our range. Some other animals that did stay put evolved from small to large and back again, as illustrated by a species of bear in Europe that changed sizes during successive ice ages, with larger bears dominating when sheets of ice covered the ground, and smaller bears dominating when the ice sheets retreated. This suggests that the species enjoyed a broad gene pool that included genes for both large and small individuals. Without this diversity, the species may have become extinct at some point during the ice age cycles.

There's an important point about these temperature fluctuations that is seldom recognized. It's part of evolutionary dogma that the single most important influence on natural selection is climate change, because it changes everything in the environment, not just the temperature. What is sometimes easily overlooked is that whatever effect climate change may have had on our evolution, we were never aware of it. Such adaptations that resulted happened without us knowing – and certainly without us consciously responding. The reason is that the changes were far too slow to be perceived. Temperature changes were typically of the order of a degree in 5,000 or many more years – far too gradual to be noticed within a few generations. The temperature changes in the tropics, where most of our evolution took place, were even smaller and less noticeable. Thus, we did not ever consciously respond to climate change, by, for example, making warm clothes, or heading north as the climate got warmer. If such things happened because of climate change, we never knew it. In the present era, the climate is warming at breakneck speed – fast enough to be noticed in a single generation. That has rarely been perceived by us before.

Though the last 11,700 years is treated as a new epoch – the Holocene–it really is nothing more than the most recent Pleistocene interglacial to be followed in due course by another glacial of the same series. There's nothing to say that the Ice Age has ended. Only time will tell, as too will it reveal how much (or little) the current interglacial warming is going to be steamed up by anthropogenic boosting of atmospheric CO_2, methane and whatnot. We don't know all the factors that drive ice ages or the cycles of colder and hotter periods within them, or indeed much about any of the planet's behavior over the really long-term. And we certainly don't know enough to suppose that we can master something as big as climate, though we can certainly mess things up enough to give it a nudge.

The Pleistocene is by far the most glamorous stage of our evolution, because it's the time in which our own genus as well as our own species evolved at a pace not seen in earlier eras. It was a time when the large animals we are familiar with today flourished, plus many other marvelous beasts that initially did well, but didn't make it to the next stage: cave bears, cave lions, giant deer, steppe wisent, mammoths; woolly rhinos, saber-toothed tigers; giant hyenas; giant ground sloths and mastodons to name some of the most celebrated. Crocodiles, lizards, turtles, pythons and other reptiles also thrived during this period. The predominantly cool, dry climates led to forest declining as more and more land became open woodland and grassy plains. This saw the extinction of many large browsing mammals, while grazers such as aurochs (nowadays cattle), buffalo, deer, antelope, rabbits, kangaroos, and carnivores such as bears, leopards, lions, dogs, cats and many more flourished. It's when we acquired many of our most remarkable attributes, qualities that enabled us to create art, to make and enjoy music, to empathize with each other, think about and discuss big and little matters, devise complex social arrangements, pursue subtle ideas. Though it's called the Stone Age, our forebears had already spent two million years banging stones together to get sharp chips, though it doesn't seem that the practise had much selective value in those days if the simplicity of the products is anything to go by. This changed during the Pleistocene, with vast numbers of ever more skillfully knapped stone blades being made right up to the replacement of stone with metals just a few thousand years ago.

Of all the things that turned us into the most crafty, domineering ape of all time by far the most influential was undoubtedly abstract thinking. We tend to take the ability for granted, but for most of our evolution the evidence supports the assumption that we behaved instinctively just as all other animals do; which is to say that we had no **conscious** concept of self; it was still unconscious. We could not 'see' ourselves behaving. Our perceptions were all responses to stimuli coming in from the outside world; we had no (or at best only rudimentary) perceptions of what we were about to do, or had done. Driven solely by instinct does not require a concomitant emotion; other animals kill or copulate without emotion; it's not essential. In fact, we too can still do these sorts of things without emotion. A prostitute can (and presumably usually does) copulate without emotion. A religious fanatic can (and often does) kill without emotion.

The evolution of emotion was part and parcel of the evolution of abstract thinking because it's a primary element—perhaps even the first evolutionary step - of the awareness of self. Instead of just acting on instinct we evolved the attachment to that instinct of an emotion that advised us (and others) of what we were about to do, or wished to do, or had done. As I mentioned in Chapter 3, the appropriate emotion is generated and managed independently of the linked instinct. It signals to us and to others what is going on in our instinctive mind. The adaptive value of this was that it helped both us and others to decide what to do in relation to the accompanying instinct, were it to be acted out. The evolution of knowing what others are thinking – what we call empathy – could have been the precursor to the full-blown capacity for abstraction i.e., to perceive ourselves and others as participants in potential future realities based on what we had already experienced and retained as memories.

That's how I see it. It's a purely hypothetical interpretation, as we don't have any concrete evidence as to how and when abstraction evolved. It remains a mystery, as does whether it was a gradual or sudden event. All we do know is that it profoundly altered the course of our subsequent evolution and is really the one adaptation above all others that enabled us to become the swarming locusts of the globe we are today.

The Pleistocene is when we can first recognize ourselves in the fossil mirror, when the first scrapbook with pictures of our funny-looking forefathers and mothers can be cobbled together. It's the date when the earliest ape of the genus *Homo*, our genus, parted the bushes with a flourish and began racing around the truly marvelous African veld alongside several species of australopiths - all no doubt barking and shrieking like baboons. This first human-like ape was *Homo habilis*, late of East Africa's Rift Valley, who was probably in the direct line that led to us. *Habilis* lived alongside our australopith ancestors for a million years between 2.5 and 1.5 Mya, when he finally went extinct. The australopiths actually survived half a million years longer than *habilis*, going extinct 1.4 Mya. Great grandfather and mother *habilis* weren't any bigger than the dinky little australopiths at 100-140 cm (3 – 4 feet) tall and 32-55 kg (70-120 lb), but they had a 35% bigger brain and more rounded skull; our brains were now relatively bigger than any other ape's. *Habilis* is credited with inventing stone knapping (though some think the australopiths took out the first patent). Whatever, *habilis* was a big user setting in motion a permanent practise that has attracted an extraordinary amount of attention from archaeologists.

I lean towards the plainest view of our lineage: that *habilis* was our Abraham, and that the next in our clade - *Homo erectus* – evolved from *habilis* to appear in the fossil record 2 Mya. This is when in the opinion of many authorities there was a dramatic bottleneck that reduced our parent species to perhaps less than 20,000 members. A combination of drift and selection further altered the bottlenecked allele frequencies, radically transforming the original animal into what became *Homo erectus*. The transformation included four interrelated complexes of changes: greatly enlarging brain size and corresponding changes in cranial anatomy; strengthening of cranial support; a shift in dental function towards more anterior tooth use and greater emphasis on grinding; and much taller bodies that doubled in weight, with many proportional body changes and changes in the visual and respiratory systems. These changes were so radical and rapid that many describe the event as a genetic revolution that laid the foundations of all our subsequent adaptations.

Around this time our jaws and teeth started getting smaller and weaker meaning that our ability to bite, tear and chew declined strongly. We also lost the sharp, protruding canines that are used in fighting. Considering that prior to the recent adoption of ground grain flours virtually all the food we ate had to be bitten, torn or chewed it's an NSDT statement that something had to compensate for reduced jaw power early in our evolution, particularly when one remembers that as foragers we had to spend most of our waking hours finding and eating food. There were three ways we compensated. Firstly, we (presumably) used our brains to become more efficient. Division of labor, the anticipation of where food was most abundant in relation to season, rainfall and local events; drying surplus meat or fish for eating later, the accumulation and cultural transmission of information; and improvements to hunting techniques must have all played a part. Our substitution of sharp stone flakes for biting and tearing was the first (fossilized) adaptation that led to or was part of the considerable reduction in jaw power and tooth size that began early in our evolution and is one of the most striking evolutionary changes we underwent. We have the smallest, feeblest jaws and teeth of any mammal of comparable size.

A slight but intriguing digression at this point is not out of place. In the beginning, when we were bipedal, social apes, steadily losing the acuity of our senses of smell and hearing, while retaining and perhaps sharpening our eyesight, though this never became as discriminating as most other diurnal mammals. This decline in the 'standard' mammal senses was accompanied by decreased muscular strength and tree-climbing aptitude. Cousin chimp is half our weight but 50% stronger in muscle power, which is why captive chimps have to be treated as dangerous animals. Whereas he retained a hand for knuckle locomotion and climbing we evolved a much more flexible hand[10] that conferred far greater dexterity at the expense of climbing ability. We adapted to walking (and perhaps jogging) and lost the physique to outpace any other mammal around our size or bigger, most notably the predators we lived with. A fast runner can touch 30 km/h – less than half the speed a lion can attain.

It has always baffled me how such a physically weedy animal as a human avoided being eaten into extinction by predators – above all lions. We've mingled with lion-like cats for all of our time in Africa, and with modern lions for at least 124,000 years[11]. Anyone familiar with them knows what bold and powerful predators they are. Present-day foragers like the Kalahari Bushmen who sleep in the open are indeed preyed upon by lions, nearly always at night while asleep. Settled Bantu people who sleep in huts are preyed upon by day when out and about, though lions will on occasion break into well-built huts to kill someone. Some lions (and tigers and leopards) acquire a taste for people

---

10 Our thumbs easily reach our pinkies, which other apes can't do because their thumbs are too short. They mostly just grasp.
11 Many lion-like fossil species go back 2 My. The evolution of the current lion is obscure, but dates back at least 124,000 Y.

and become notorious man-eaters. After millions of years of evolution we are still lion fodder. So what is it that our pre-urban ancestors, Kalahari Bushmen and like people, not to mention *habilis* and *erectus* did to keep lions from wiping them out rather than just snack on them sporadically.

One has to remember that until about the time our brain started enlarging (i.e., for most of the Pleistocene) we apparently didn't have spears or bows and arrows, though we can't be sure that we didn't have wooden spears that never fossilized. That meant that as we lost our physical strength and speed we lost the ability to dodge a lion, without, apparently, acquiring anything that would frighten or seriously injure a lion – or even a leopard or hyena or wild dog. There were not always really big, climbable trees handy. *Erectus* was as tall as we are and not anatomically built for climbing; nor is it thought they slept in tree nests. Fire was no protection - lions aren't put off by it.

There is an intriguing possibility. Some present day Bushmen claim they and lions have a 'mutual understanding' not to mess with each other. What that means to me is that a given band of early humans had a territory, as do lions. Where these turfs overlapped the people and lions would have got to know each other by sight. Provided the humans never acted like conventional prey by running away, but (without attacking) stood their ground aggressively, the two parties could have habituated to each other. It's quite a common observation that potential prey living in a predator's vicinity come to be ignored; but a strange prey animal of the same species that suddenly appears is immediately pursued.

There are other aspects to consider. Most predators operate so as to get the drop on their prey, and if seen before they are ready to kill will usually abandon their hunt. Leopards mobbed by monkeys will make off. Predators can be thwarted by prey animals grouping tightly, making it difficult for an individual to be singled out. All these considerations are only relevant during daytime; they fade away with the setting sun. The lion roars to remind the zebra that he is afraid.

Even after the evolution of spears and powerful bows and arrows, it's still hard to imagine efficient repulsion of packs of large predators like hyenas and lions. Animals like buffalo are vastly stronger and better armed, and will stand and fight attacking lions. But, it doesn't deter the cats; a lot of adult buffalo die at the claws of lions.

One inevitably thinks of shelters in the context of predator defense. Other Great Apes don't make shelters, but they live in lion and hyena-free forests. Their largest predator – leopards – can only kill chimp-sized apes and can be over-powered by a gorilla or orangutan. It's notable that cousin chimp usually sleeps aloft – presumably to foil predators – as do orangutans. In the many animals that do make shelters (burrows) they are primarily used at times of extreme vulnerability, i.e., when asleep or raising offspring. I assume that the same applied to us, i.e., shelters reduced both infant and adult mortality. Their use, together with *bomas*[12] could've been widespread, but with scant chance of fossilizing would've left no trace. One must always remember that while most human fossils have been found in caves, most people lived out in the bushveld along rivers and near waterholes where fossilization is less likely to occur. I find it hard not to assume that the making of shelter in Africa occurred quite early on in our evolution primarily driven by the need to protect ourselves from predators.

Stone hut foundations in Europe go back over 400,000 years, when construction was already advanced with substantial stone foundations; the first shelters must've been much older. Whether or not we were building huts in Africa at that time is unknown. We used caves where they existed, but in much of the African savanna we would have had to make shelter. There's no archaeological evidence for this, neither would one expect it where shelter was of plant material. And shelter would have to have been strong to actually prevent predation. But, a comparatively flimsy shelter could have served to consolidate a group of people and facilitate a coordinated repulsion of marauders. All speculative, of course. It may be that we always did lose a considerable number of individuals to predators – to me it seems highly likely, whether or not we made shelters.

To continue the narrative. Secondly, we made tools – digging sticks for roots and tubers, artificial chewing by pounding tough plant foods, stone blades to replace biting and tearing, and spears, stones and clubs for hunting. At some unknown point we made snares and other traps.

Thirdly, cooking. This was the next archaeologically visible adaptation to the weakening of our jaws. This must have begun with meat because apart from tubers our ancestors couldn't cook plant foods without containers for boiling. We can't say when fire was first used to cook meat but there's some evidence of it 1 Mya in South Africa, and by 400,000 Ya stone hearths were in use. Likewise, we can't determine when other improvements in finding and preparing food evolved, but given the early reduction in jaw size there must have been many improvements early in our evolution. This assumption is partly based on the exceptionally long period of human childhood during

---

12 Boma, a rough enclosure of cut vegetation (Swahili)

which all the difficulties of chewing and digesting food are considerably greater than in adults.

The selective advantage of cooking is that it makes food softer and more digestible. The digestion of raw meat for example consumes a third of the meat's energy. Cooking reduces this by 20%, meaning that a given amount of meat provides much more energy cooked than raw. Against this is the loss of fat if cooked over flames. We may have cooked unskinned carcasses or limbs to minimize fat loss – nobody knows, or is likely to find out such things. The same goes for cooking plant foods. Without boiling water it must be supposed that plant foods could only have been cooked under embers by caking in mud or wrapping in leaves. We're unlikely to ever know; as such actions don't fossilize. Although ceramic pots weren't invented until 20,000 Ya in China, and only 10,000 Ya in Mesopotamia, it's possible our forebears boiled vegetables in skin containers long before hard pots. Water can be boiled in flammable containers provided the container material is kept in contact with the water; but how long ago we discovered that fact is quite unknown and unguessable (and not that widely known even today). All that can be said for sure is that we've been eating progressively more chewable and digestible plants and animals ever since our jaws started weakening.

Many of the most striking features of our anatomy compared with our fellow apes are generally considered to be strongly associated with the trend towards making food more available, easier to eat and more digestible. Firstly, our brains got larger as our bodies got weaker. Our jaws and teeth in particular steadily got smaller and punier. Our present-day gut is simply too short to cope with an all-raw diet as it's only two thirds the size of a foraging ape's of similar weight. In other words, we evolved smaller and smaller guts (which diverted less and less energy to digestion) as we adapted to better-prepared, higher energy foods. The physical weakening of our bodies and jaws might at first seem at odds with the most fundamental of all necessities – getting food. The best explanation seems to be that our evolution selected for reducing the bodily consumption of energy in favor of enlarging our brain. The bigger brain compensated by enabling more efficient methods of procuring and processing food, as well as other survival adaptations. At the same time, the enlarging brain consumed more energy than any other tissue, weight for weight, until today it consumes up to 23% of our total energy use, though it's only 2% of total body tissue. The net evolutionary outcome was to reduce overall bodily energy consumption sufficiently to enable a larger, high-energy brain.

Another adaptation to making feeding more efficient was the social one of grandmothering. A unique feature of our reproduction is the cessation of ovulation (menopause) quite early in life. The evidence suggests that many of the women who survived until menopause lived for many years after, raising the question of why selection seems to have favored this. It's thought that such women enhanced the survival of younger women's children by easing the burden of feeding them after weaning, i.e., indirectly increasing fecundity. Such is the case among contemporary hunter-gatherer and subsistence farmers and is likely therefore to have evolved a long time ago.

We have several other physiological peculiarities that are unique among our mammal contemporaries, which can be seen as helping make way for a huge, energy-costly brain. Along with the decline in muscle power went the loss of body hair (except the hairy 'hat' over the head, and armpits and pubes) and extreme thinning of the skin, that both reduced bodily energy consumption and retention of heat by insulation. These efficiencies became selectively important owing to an inevitable consequence of the decline in our senses of sight, hearing and smell. For the decline forced us to become entirely diurnal for foraging and other physical activity; on dark nights we could only blunder around at best. And being diurnal meant that we had to be at our most physically active while the sun was up. This put a selective premium on keeping cool – especially keeping the all-important brain cool. This in the opinion of many (and I am one) explains the origin of our unusual cooling system.

We sweat for evaporative cooling more than any other mammal - up to several liters of water daily, mostly from the upper body, the part most exposed in bipedals to the wind. The brain has an extremely high metabolic rate, emitting up to 23 W of heat (which is 23% of the total from only 2% of the tissue). Yet for shock protection it's encased in a thick, well-insulated layer of bone and is therefore hard to cool except via its blood supply. Cooling the blood while it's in the body is a solution - analogous to cooling a car's engine via a remote radiator. Whatever the reason for our high sweat-water consumption it will have forced us to live close to water throughout most of our evolution, given that as far as we know we had no means of carrying or storing large quantities of water until the recent evolution of pottery. This water dependence would have kept our ancestors out of vast tracts of African bushveld during their respective dry seasons, and probably led to relative crowding around water at such times with concomitant social consequences.

After more than a million years of evolution in Africa *erectus* was granted the title of our Last Common Ancestor (LCA), who spawned us and the three varieties of us who lived contemporaneously for the next few hundred thousand years. This much-argued-about LCA is dated to 800,000 Ya and as

Timeline 2   Evolution of brain and body size

	Average brain volume(cm^3)	Individual variation(cm^3)	Height (cm)	Weight (kg)
Chimpanzees	400	300-500	120-150	40-70
Australopiths	470	390-545	100-140	31-50
H. habilis	630	509-752	100-140	32-55
H. erectus	1000	850-1100	145-185	40-68
H. heidelbergensis	1200		170-180*	
H. neanderthalensis	1500	1250-1760	150-180	64-82
Early humans	1350		Variable	Variable
Modern humans	1400	900-1880	163-176	60-80

* Tentative as only 3 specimens provide heights

far as the fossil record currently goes is most likely to have been *H. erectus* – not that all concur. Many prefer to call our LCA *H. antecessor*, distinguishing it from *erectus*. A consensus among archaeologists and all those who study such things about any aspect of our evolution is and probably always will be as likely as lions agreeing to become vegans. That the LCA was *erectus* or *antecessor* does seem likely, though, because many *erectus* fossils have been found all over Africa and Eurasia, suggesting that if there was some other species of the same era that gave rise to us it should have had the same chances of fossilizing. I should mention that some fossil fogeys recognize about a dozen additional species of *Homo* based on differing interpretations of the same fossil record – it's all a matter of opinion and may never be settled in favor of any one view. Recent finds in Israel and China suggest that the received evolutionary sequence might have to be revised. It's a fascinating subject and each new fossil discovery sets everyone off again promoting and discounting this that and the other. The broad outline is not in doubt, and thankfully the details don't affect this enquiry, however absorbing they may be.

The LCA is important because it marks the point, 800,000 Ya, when we and the progenitor - *H. heidelbergensis* - of our two closest cousins, *H. neanderthalensis* and the shadowy Denisovans came into existence as distinct species. The Denisovans are a mysterious bunch only known, paradoxically, from a genome deciphered from a rather tatty smidgeon of DNA fished out of a finger bone in Siberia. Only two other small Denisovan bones have been

Timeline 3.   Notable Old Stone Age events from the start of the Pleistocene to the singular point when our already large brains started enlarging even more.

Mya	Event
550,000 Ya	Neanderthal ancestors evolve from heidelbergiens in Eurasia
600,000 Ya	*H. sapiens*, and later, Neanderthal brains start enlarging
700,000 Ya	Homo Heidelbergensis evolves in Africa
800,000 Ya	LCA of *Homo sapiens*, thought to be *H. erectus*
1	Many non-ruminants go extinct as climate dries more strongly with successive cold, dry pulses. Grassland had been increasing for last 4.5 my
1	Definitive *H. erectus* enters Asia (but not yet Europe) from Africa
1	Loss of body hair complete, but for head, armpits and pubes
1.2	Black skin evolves
1.4	Last australopiths go extinct
1.5	Acheulian (hand axe) period begins. Also first butcher marks on large animal bones. Chimps & bonobos diverge
1.8	A branch of small erectus-like people (*Homo erectus georgicus*) reach Europe
2	Severe bottleneck culminates in replacement of *H. habilis* by *H. erectus*. Probable use of speech. Possible first meat cooking in Southern Africa
2.5	*Homo habilis*, first of our genus *Homo*, evolves in East Africa's Rift Valley from one of the many australopiths. Made hand axes.
2.6	End of the Pliocene, start of the Pleistocene -the Stone Age, the Ice Age

found so that we have no idea what this cousin looked like. Probably they were doppelgangers for Neanderthals whose many fossils may include some that are actually Denisovan, but indistinguishable without DNA; otherwise they should have left as many fossils as we and the Neanderthals did. Nobody actually knows. Even Dick Tracy would be scratching his head over that one.

H. erectus was a remarkably hardy, footloose species who found his way into Eurasia 815,000 Ya and soon spread widely. In Africa, *Homo heidelbergensis* evolved 700,000 Ya and by 500,000 Ya some had made their way into Europe. These European heidelbergians gave rise to the famous Neanderthals and apocryphal Denisovans, while the African heidelbergians gave rise to us, *H. sapiens*, (and Southern Africa's *H. naledi*) about 300,000 Ya.

We, *H. sapiens*, had evolved into the size and shape we are today (except for our craniums) by 200 Kya. Unlike all our contemporaries we stayed in Africa until quite recently, and while there's evidence of several abortive safaris into the Middle East we only established ourselves there permanently about 60,000 years ago; crossing the Sinai was evidently not all jam & Jerusalem. We kept to the warmer Middle East and southern parts of Asia for a long time before invading Europe 45,000 Ya.

H. erectus was a tough cookie who survived until just the other day in evolutionary terms - a mere 70,000 Ya (or as little as 27,000 Ya according to some). Some *H. erectus* were the same size as us. They had a much bigger brain than *habilis*, though it never grew as large as ours. He's thought to be the first human-like ape to use fire.

A word about the terminology used when talking about us. It's impossible not to use the epithet 'human', or the truly dreadful 'human being,' when discussing our evolution, even though it has almost sinister undertones. The word human itself is harmless, but its purpose isn't. It was invented to set us apart from Nature and every time one uses it one reinforces its underlying principle of apartheid; the assumption of a superior race elevating itself above an inferior horde of savages who must look after themselves and observe social distancing at all times. The addition of 'being' – which even respectable scientists use in a formal context – doubly repels the notion that we are animals. It masks the facts of life, which apply to humans as much and in the same manner as they do to all other animals. We are all of us of the same ilk, the product of the same Nature. But, the presumption of our inherent 'superior difference ' is so deeply imbedded in our narcissistic self-awareness that trying not to use the 'human' word is doomed from the outset. It's a shame that hysterical perceptions of chauvinism turned the perfectly harmless 'man' to describe our species into a dirty word; it's a much better epithet than human.

Archaeologists like to partition the Stone Age – Paleolithic as they call it - into three periods: the Lower Paleolithic, 3.3 Mya–300 Kya; the Middle Paleolithic,300-40 Kya; and the Upper Paleolithic,40–10 Kya (the period in which the Fertile Crescent played such a crucial part).

The 'official' end of the Pleistocene coincides with two striking events: the beginning of serious farming that, for us, brought the curtain down on the Good Old Days, the colorful Stone Age that had fostered us for nearly 3 million years; and the extinction of more than three-quarters of the large ice age animals, including such prodigious beasts as woolly mammoths, mastodons, saber-toothed tigers, woolly rhinos, giant bears, giant ground sloths and many others. It has long been fashionable to blame predation by us as the cause – the so-called Pleistocene Overkill. Others posit a sudden warming event 13,500 Ya, some the onset of the current interglacial warming; and yet others the explosion of a three-mile-wide comet over Canada. Nobody actually knows; but the Pleistocene Overkill is to me pure horse feathers.

First of all, we had lived together for hundreds of thousands of Pleistocene years, yet all of a sudden, just before the end of the Pleistocene, we allegedly speared the lot. It's worth noting that after we allegedly slew all those Pleistocene Mighty Ones (and not just lots of them but *all* of them) huge populations of other large mammals built up -elephants, hippos, bison, buffalo, rhinos, and so on. Up to around 300 years ago, we could not eliminate these large animals– as evidenced by their abundant status in Africa and North America up to the time of arrival of Europeans and guns. With an estimated average world population of one million people in the Pleistocene and only spears in our hands how is it we were able as so many want to believe to exterminate around a billion individuals of 50 species of large animal? Pre-gun man of the present era did not commonly hunt large, fierce animals like elephants or buffalo with spears or bows and arrows; he mostly did so with poison-arrows, because the wounds inflicted without poison simply were not crippling or fatal, not to mention the danger of taking on such beasts at point blank range. Moreover, hunting with poison-arrows in Africa was dangerous and only practised by a small number of doughty hunters. Even after the arrival of Europeans and guns in Africa, the extermination of large African animals is only being achieved by the destruction of their habitat and replacement by cultivation; guns only turbo-charge the underlying process.

Screeds have been written about what we all spent our time doing in the Pleistocene, but, as usual, we have at best only a hazy notion. As can be seen from Timeline 3 we began to talk, became hairless except for the head, armpits and pubes, evolved black skin, improved our knapping, hunted large animals, cooked meat, made huts. Which species was doing what and how we related to one another is almost entirely speculative because by far the greater part of our activities left no fossil traces; all we can be sure of is that life became infinitely more complex than hitherto.

****

Well before the end of the Pleistocene something truly remarkable occurred that profoundly changed our behavior, culminating in us abandoning forever the foraging way of life we'd pursued ever since our tails dropped off and we clambered down from the trees: our brain started to get bigger. By the end of that process it was 17% bigger. And presumably better – depending of course on what one deems better. Much better at surmounting Nature, which meant eliminating those parts of Nature that got in our way. And most of Nature *was* in the way. Better or not, it was certainly quite different.

Part 4

600,000 Ya – 35,000 Ya

Chapters 7 and 8

Chapter 7   The Big Beautiful Brain

Spurt in brain growth starting 600,000 Ya – surged 200,000 Ya, as we 'officially' became humans – 17% overall increase, all to prefrontal cortex, parietal lobes and cerebellum, i.e., the 'thinking' parts – radiation into the last four species – possible evolution of abstract thinking during this time – many 'civilized' developments, better spears, knapping, grind stones, stone hearths, stone huts – first signs of advanced inclusive fitness, care of cripples, abstract designs, body painting – first sign of evolution of instinct for belief in afterlife.

At this point in our evolution, 2.5 million years after we waved our dinky little australopith fore-parents goodbye and strode off into the East African sunset in search of greater wisdom and happiness, we already had the biggest brain relative to body size that the earth had ever seen in any animal. During the evolution of *erectus* from *habilis* it had enlarged faster than it ever had before (or since) - about a gram every 2700 years. For the next million or so years until the evolution of *Homo heidelbergensis* its growth slowed to a gram every 5000 years or so. Then, 600,000 Ya it suddenly began growing bigger much faster again –averaging about a gram every 3000 years – almost as fast as it had during *erectus'* heyday. There was a difference, though; a big difference. The first period of rapid growth – that of *erectus* - was accompanied by a considerable increase in body size, which was bound to cause a corresponding increase in brain size. The second growth spurt was to the brain only – body size didn't change. That difference makes it by far the fastest increase in brain size of our entire evolution. Something - or things - really profound must have been involved. The size increase lasted for the next 565,000 years, with a surge 200,000 Ya – just when we officially became the humans we are today. The increase ended about 35,000 years ago, since when it actually became slightly smaller. And that's the size it is today, averaging about 1400 cm^3 in volume and 1.5 kg in weight – The Big Beautiful Brain.

In case a gram of brain every 3000 years doesn't seem that much to you, bear in mind that it represents a total increase of some **15 billion neurons**, mostly to the thinking part of our brains; that's a tremendous amount of additional brain power.

By the end of the Big Braininess our brains were strikingly globular with steep frontal, bulging parietal, and enlarged, rounded cerebellar areas. The cerebellum is involved in spatial processing, memory, language, social understanding and emotion. The parietal bulging was actually caused by the enlargement of a deep-seated area that includes the *precuneus* – an important hub of brain function. Our faces also became smaller and more retracted - a shape that has stayed unchanged to the present day. These anatomical changes did not happen to Neanderthals, from which we must suppose that our behavior also changed in comparison with them. Although his brain enlarged as much as ours its growth was allometric, i.e., driven by the relative growth rates of all its components; ours was not allometric, being driven only by the specific forces of the brain parts that were enlarging. That's why Neanderthal skulls are not globular, but more elongated than ours.

Though brain size in animals correlates closely with body size, as would be expected since more tissue demands more nerves, more processing, our brains are far larger than our body size requires. Much bigger in fact than those of any other animal of similar size, except for cousin Neanderthal whose brain was as big as ours, though differing in physically small but behaviorally influential ways (As we lack a Denisovan skull this ghostly cousin's brain size is unknown). The 17% size increase meant a corresponding increase in the brain's energy consumption, which meant finding and eating either a lot more food, or more energetic food – such as grass seeds.

Selection for a bigger brain must have been driven by unusually powerful selection pressures because it imposed a considerable metabolic penalty as well as forcing substantial anatomical changes to the skull. It's a logical assumption that those selection pressures must have been primarily mental as, apart from the skull, there were negligible anatomical changes over the period. In other words, our behavior must have changed. As behavior can't fossilize we can only draw inferences from the smidgeons of bone and other things contained in the fossil record.

One of the more visible changes at the beginning of the Big Braininess was the radiation of our genus into the four species (including us, *Homo sapiens*) that coexisted along with *erectus* until near the end of the enlargement, 35,000

Ya. The first was *Homo heidelbergensis*[13], some of whom trekked out of Africa into Eurasia around 600,000 Ya where they became the progenitors of the famous Neanderthals. About 350,000 Ya a group of Neanderthals in Asia split off to become the Denisovans, whom we ultimately met when we wandered into Southeast Asia 60,000 Ya. When we got into Melanesia and Australia we took some Denisovan genes with us. Contemporary Africans do not have any[14] Neanderthal or Denisovan genes - a reflection of the fact that initially we and some heidelbergians stayed in Africa, while Neanderthals and Denisovans never moved out of Eurasia into Africa. The African contingent of heidelbergians evolved into us and by about 350,000 Ya we had evolved the main anatomical features that characterize us today. It wasn't until about 92,000 Ya that we seriously invaded Eurasia for good, where we met up for the first time with our Neanderthal and Denisovan relatives who'd already been pottering about there for the last 300,000 years. All three of us interbred occasionally, but whether forcefully or consensually we can't say. Through all this tough old *erectus* was still with us, but his brain, though it enlarged a bit, never saw the increase ours and the others' did. He went extinct in Africa at least 500,000 Ya after giving rise to *Homo heidelbergensis*, but flourished in Eurasia alongside the Neanderthals and Denisovans. For reasons unknown, but before our arrival, the fortunes of all three declined until by 100,000 Ya they were all dancing their Last Tangos in Paradise. The strange thing is that it seems none of us ever hooked up with *erectus* in spite of our wide distributions. We were all hunter-gatherers so must all have had our eyes on the best bananas; but never, it appears, simultaneously. We must have bumped into each other from time to time, but there are no *erectus* fossils at the same sites as any of the other species of *Homo*. One or both of us clearly didn't like the look of the other.

The puzzle as to when we evolved abstract thinking is inevitably evoked when discussing the Big Braininess. That it began prior to the Big Braininess is certainly possible, though there's no unequivocal evidence of that. That it matured during the time of brain enlargement is certain. Apart from the cranial changes such archaeological evidence as exists strongly suggests that this half million years of rapid brain enlargement was when our 'civilizing' mental attributes evolved, though we can't say exactly when or in what order each attribute became genetically fixed. A few do stand out, though, and one

---

[13] Many classify the first as *H ergaster*, which gave rise to *H heidelbergensis* after leaving Africa for Europe. I'm following those who regard *ergaster* and *heidelbergensis* as the same species.
[14] North Africans do carry some Neanderthal genes thought to have been relatively recently introduced by returning *H sapiens*.

is better spears. Within a hundred thousand years of the start of the Big Brain spurt heidelbergians southern Africa began hafting stone spearheads to their wooden spears. A given spear's thrust, or throw, now made a wider, deeper, more severe wound – to me, a big advantage given how difficult it is to hunt and fight efficiently with hand weapons. Seemingly small advances in design can make big differences. The contemporary experience with ever-changing weapon design underlines the considerable advantages of more lethal as well as efficient devices. It's hard to believe that we didn't haft stone spearheads as well, though the fossil evidence suggests that we and Neanderthals used only plain wooden spears until more recently, about 300,000 Ya.

Another feature of our brain expansion is that it was accompanied by more skillful knapping to improve the so-called Acheulian tools. There were two features to this. One was refinements of existing types that by the end of the Stone Age resulted in beautifully proportioned, symmetrical tools that suggest the incorporation of an aesthetic element. Observations on contemporary people learning to make these tools show that knapping involves the prefrontal cortex, not just motor control. The prefrontal cortex is associated with higher-level thinking and reflects the need to abstract what must be done during knapping to achieve a predetermined object.

The other feature was the so-called Levallois knapping 300,000 Ya whereby partially knapped stone cores were later finished into blades. It's supposed that these unfinished cores were produced by Neanderthals specializing in the practise who traded them to others who finished them in the shapes they wanted.

Although most of these tools are called hand axes there's no evidence that any of the Stone Age artifacts were in fact axes, i.e., heavy blades for chopping wood, etc. True axes are found at few sites, and only date back to the end of the Stone Age. It seems that the Acheulian hand 'axes' were mostly cutting, flensing and scraping tools (uses that any one of them could be put to), as well as points for spears and arrows. I find it odd that while archaeologists describe similar-looking stone flakes by a wide range of use-based names – even including 'meat cleavers' – they never call them knives. To me, that's what they are: knives – the universal tool we still use for all the manual functions that involve some sort of cutting. Whatever, they were certainly important as the so-called hand axe dates back one and a half million years, and cruder tools go back to the time of the australopiths. They are a continuous accompaniment of human evolution and there can be no doubt that they set the neural scene for the huge variety of dexterous actions displayed by contemporary man. Nevertheless, it seems to me that the durability of stone gives an undue prominence to the significance of stone tools. While it's interesting to watch cousin chimp crack a nut with a rock, and it would certainly be interesting to know what we got up to with the ubiquitous hand 'axes' our ancestors obsessively made so many millions of, it's infinitely more interesting to learn what we apes did **without** the help of bits of rock. But, all we've got are the forensic inferences drawn from a contemplation of various artifacts associated with fossil bones that give at best only a hazy idea of the cultural changes that took place.

To continue. By 400,000 Ya stone hearths for cooking were widespread in Europe and by 280,000 Ya grind stones appear alongside hearths. The oldest bedding made of vegetation so far discovered is 200,000 Ya. It's at this point – 200,000 Ya – that a second burst in the overall Big Braininess took place. For most people who study our evolution this is a magical date, because it marks the point at which we must have undergone one of the most influential – perhaps *the* most influential – mutation or mutations to our intellectual apparatus. Not only that, but it also marks the point at which the final anatomical changes that most authorities regard as defining us as a species took place. Geneticists have found that the genes that are unique to us are mostly associated with brain development and function. From this point on all human fossils are true *Homo sapiens sapiens*. Although our brains continued increasing in size the amount was small and did not substantially change our cranial characteristics.

What did change during this final phase of the Big Braininess was our behavior (see Timeline 4). The first archaeological signs of many of our most advanced activities fall into this 165,000-year period from when we became, anatomically, what we are today to the end of the Big Braininess, 35,000 Ya. By far the most striking of these behavioral changes are those relating to our social behavior within our groups; in a word, our so-called inclusive fitness adaptations.

A standout inclusive fitness adaptation is care of crippled individuals. The 92,000-year-old fossil of a severely brain-damaged human child, who died aged 12, some four years after the brain trauma, was found in Qafzeh Cave, Israel. The protracted care of this child implies that such 'altruistic' behavior was well established and had presumably begun long before this date. The significance (to me) is the implication that we were capable of strong affectionate bonds among kin on top of the instinct to look out for each other; by this point we had become emotional animals. That's just whimsy, of course, we don't know when we evolved our complex range of emotions; but quite possible.

Timeline 4.   The principal known events during the enlargement of our brain.

Kya	Event
35,000	Brain stops enlarging. Neanderthals virtually extinct
36,000	Dogs possibly associated with humans.
37,000	First cold Heinrich event
40,000	Dogs evolved from now extinct wolf species in SE Asia.
40,000	End of Middle Paleolithic, start of Upper Palaeolithic (late Stone Age), ends 11,500 Ya. Neanderthals make bone fipple flute, i.e., music well established. Also associated with first ivory, bone, antler & shell tools. Grinding stones either for seeds or pigments (or both).
46,000	First modern humans enter Europe from Eurasia
48,000	Neanderthals heat-treating birch to make pitch for hafting.
60,000	Neanderthal symbolic cave drawings (at least long-lasting ones). Oldest Neanderthal hyoid bone similar to human's i.e., they probably had speech.
65,000	*H. sapiens* reaches Australia. Did not bring bows & arrows, but must've brought knapping, speech, burial, rhythm, singing, spears, atlatls, fire & body painting.
68,000	*H. sapiens* reaches Indonesia.
70,000	Oldest known arrow heads, Sibudu Cave, SA., made of bone & stone
74,000	Second bottleneck. Some say only 20 k humans survived Toba volcano eruption.
92,000	Care of cripple, Qafzeh Cave, Israel. 15 other burials associated with pierced shell ornaments, ochre & manganese colouring.
92,000	First *H. sapiens* In Eurasia, i.e., first successful safari out of Africa. Making fishing tools. Neanderthals had been there for last 400,000 years.
100,000	Symbolic material culture. Engraved ostrich eggshells in South Africa. Abstract designs. Increasing use of red ochre in Africa up to 50 Ka.
112,000	*H. erectus* nearing extinction in Eurasia – last known fossil, Java
120,000	Temperature 2 degrees higher than today
130,000	Neanderthals using eagle talons for jewellery
130,000	Earliest certain Neanderthal burials Mt. Carmel and the Galilee, 40 individuals
130,000	Savannas shrinking, forest expanding, i.e., hot & wet
170,000	Clothes worn routinely by us and Neanderthals
200,000	Oldest true humans, *H. sapiens sapiens*. Spurt in brain enlargement.
200,000	Bedding found in South African cave used by humans
250,000	*Homo naledi* evolves in South Africa
280,000	More complex stone blades. First grind stones
300,000	End of Lower Paleolithic, start of Middle Palaeolithic. Neanderthals & Denisovans diverge, probably in Middle-East.
300,000	Neanderthals hafting stone spear points in Europe. Evolved Levallois knapping whereby partially knapped stone cores were later finished into blades.
350,000	Anatomically modern (but for brain size) humans evolved from heidelbergians

	in Africa, but not yet definitive H. sapiens
400,000	Permanent, widespread occupation of Europe by Neanderthals. Using long wooden throwing spears (2.1-2.4 m x 3-4.7 cm), stone hearths, cooking.
424,000	Many stone hut foundations 3.5x3 m in Europe
500,000	Heidelbergians hafting stone spear points S. Africa
500,000	*H. erectus* extinct in Africa, but spreading over Eurasia
550,000	*Homo neanderthalensis* ancestor evolves (in Eurasia) from H. heidelbergensis
600,000	*Homo sapiens* brains start enlarging,

Inferences from the evolution of body lice suggest that some Neanderthals at least were full-time wearers of clothes by 170,000 Ya. Other evidence of abstract thought is the engraving of abstract designs on ostrich eggshells, and the use of red ochre, presumably for body painting, that was practised in Africa at least 100,000 years ago. The 'paint' was made by pulverizing iron-rich red ochre rock with charcoal and fat.

As can be seen from Timeline 4 many other adaptations occurred in the final phase of the Big Braininess. The bow and arrow evolved in southern Africa, Neanderthals were making music, and it's highly likely that we did too, though there's no archaeological evidence for it. We were almost certainly using red ochre for body adornment, and a variety of objects was being made into jewelry.

It seems clear that the evolutionary pathway of our enlarging brain could only have been towards, for want of a better word, cleverness. We came to rely more and more on brainpower instead of muscle-power. It's self-evident that contemporary man copes with basic instinct by highly labile cultural brakes, switches, and shunts that circumvent the slowness of gene evolution by being inherited verbally.

Furthermore, over the period that our brain enlarged - the Pleistocene — the environment repeatedly cycled drastically from hotter to colder and back to hotter climates, which would have called for an improbable matching selection back and forth of those genes conferring adaptation to climate change and all that goes with it. We rode the changes rather than submitted to them. The accidents of mutation and selection that resulted in our ability not just to manage but to flourish in this volatile environment must have been influenced by the genome we started with that defined the subsequent windows of evolutionary opportunity that happened to be selected. We don't know the details of this genome, but we do know at least a little of what we were like physically and can speculate about the nature of the evolutionary pathways that underlaid the all-important changes to our enlarging brain without being overly fanciful.

****

To me, one of the most profound developments during this period was burial of the dead, which is associated with the evolution of belief in an afterlife that I discuss in Chapter 8 below. The oldest known unequivocal burials are in the Middle East dating back 130,000 years, showing that burial was a well-established practise by that time. The belief in an afterlife led to religion, which has had and continues to have a profound effect on the shape of civilization.

Part 5
600,000 Ya – 35,000 Ya

Chapter 8   Belief and the Supernatural

Belief without evidence is what is told by one who speaks without knowledge of things without parallel.
Ambrose Bierce

Evolution of the instinct for belief - belief in supernatural afterlife with burial of dead – belief in supernatural world to explain otherwise inexplicable events such as death – magic – evolution of religion out of supernatural – use of religion to unify and coerce – belief in science as alternative – rhythm and music – evolution of our conscience and its enforcer, guilt – morals.

One of the strangest things about our evolution is our belief in the supernatural. Looked at from the perspective of a present-day, urban, educated person the whole idea of the supernatural can seem ridiculous, a spurious departure from a rational perception of the world. That much of the rational world remains unknown or unexplained is readily taken by such educated urban people as nothing more than a lack of knowledge, or inability to perceive what our sense organs didn't evolve to respond to. In other words, not inexplicable; just waiting to be explained. Bizarre though this weird behavior may seem to many of us today we should remember that until just a short time ago **everyone** took the supernatural for granted, be it in the form of animism or gods. It was not something one could choose to adopt because society offered no alternative concept, or any other way of looking at the world. The supernatural **was** so; everyone knew that, because everyone was taught it. Above all, it was, ironically, reassuring, because it explained everything that would otherwise have been inexplicable and therefore alarming. The belief had enormous influence over what society did - and continues to do; we are not yet ready to abandon this way of responding to our capacity for belief.

Understanding the mental determinants of belief in a supernatural world full of imaginary forces has exercised thinkers throughout history and continues to be debated, or rather argued. I think the continuing difficulty arises because of two, related things: firstly, even today not everyone accepts that all our behavior is initiated by gene-driven instincts modulated by cultural controls; and secondly, with belief we are dealing with a deep-seated, ancient instinct lacking the clear-cut, easily perceived features of plainer, more obvious instincts. Some may assert that believing in something is not instinctive at all, but culturally generated; but its universality argues against that. A complication is that the belief instinct is not an operational pre-formed behavior like eating or fleeing danger, but one of several that is inherited as a template, to be filled in by the cultural hand-me-downs of the particular time and place. Like the instinct to learn language, a belief will only develop in the presence of the right cultural inputs. A congenitally deaf person cannot spontaneously learn to speak; the template requires the learner to hear others speaking.

For my part, I find it easy to understand our readiness to believe in the supernatural by reference to the undoubted existence of the extremely strong need for the conscious mind to understand everything it perceives. We need an explanation for everything. We suffer anxiety until there's closure to an unsolved accident or

> Apes are apes, though clothed in scarlet.
> 'Twas only fear first in the world made gods.
> **Ben Jonson**

murder. Anything that happens to us must be comprehensible and complete. Unfinished business keeps one awake at night. This pressing need for clarity, completeness, closure, explanation – call it what you will - in itself seems to me to be a NSDT statement, because the mechanism that makes the final decision over everything we do– our conscious mind – must have unambiguous, explicit perceptions of everything it's reacting to. For it to make decisions not based on firm (preferably accurate, but not necessarily so) perceptions of what's going on around us is to act irrationally. Shooting in the dark. That's not maximizing the individual's security. On the contrary, it's exposing him to danger. He who hesitates is lost. We must know what we're doing and why.

But, and it's a huge but, not everything the early mind (not to mention a huge fraction of the contemporary mind) perceived was explicable in material or familiar terms. When we lived more or less as any other wild animal, as we did until just the other day in evolutionary terms, what happened in our environment was keenly watched. Extremely closely, as can be seen in any wild animal. It was too dangerous not to, or to simply ignore something. There had to be a reaction to everything, because everything had to have an explanation before it could be ranked safe or dangerous and treated accordingly. Most things generated an innate response, but from the beginning, we witnessed many palpable but inexplicable physical things- earthquakes, whirlwinds, thunder and lightning, sunsets and sunrises, volcanoes, diseases, etc. Even today, many people in cultures close to Nature who are uneducated about what's involved adduce magical explanations for such phenomena; mythology is full of it. We were well aware that powerful forces were at work – they could often be felt, or the aftermath seen. Since these forces didn't match anything we were familiar with they were by definition invisible. As I see it, there was no other way of coping with them than to suppose and then to believe that they emanated from another, invisible, intangible world alongside the perceived world of everyday, comprehensible things; what came to be universally known as the supernatural. To explain mysterious events as caused by intangible agencies that did things we couldn't do seems to me a quite small step that had strong selective value. It also seems natural that these agencies should have been modeled on us, both in form and behavior, differing only in their invisibility and power. We call them spirits or gods or demons, and once believed in there was no limit to either their form or power since no one could rule any of them out. Which is why there has been such a vast horde of fairies, gods, spirits, ghosts, devils and whatnot conjured up by the wise ape everywhere he went.

It's an NSDT statement that superstition, the belief in magical forces, presupposed the capacity for abstraction. Abstract thinking had to come first. Up to that point we and other animals could only perceive the present based entirely on the material objects we were sensitive to. The present was constantly changing, but only by reference to what we could physically sense. What changed was that we acquired the ability to perceive scenarios that were entirely mental, i.e., were not stimulated by anything external. Their stimulation was wholly from within the mind. They were essentially recreations of past experiences that were more than simple memories; they had the quality of resembling active, current experiences, which could lead to other current experiences even though nothing was actually happening in reality. This ability of the conscious to think in the abstract used the same neural pathways as were used to perceive the present based on memory and external stimulation. What changed was the substitution of internal for external stimuli to create an internal reality based on imaginary scenarios that was independent of the actual present in terms of time.

One can see this as either a monumental, isolated and novel evolution or the result of a series of small changes to existing neural mechanisms. That it had a monumental influence on our subsequent evolution there's no doubt; but how it came into being is not self-evident. That it was associated with the period of brain enlargement is to me certain, if only because something truly monumental must have been responsible for such an extraordinary event to have been selected for. I subscribe to the view that it was a gradual evolution of existing neural mechanisms for two main reasons: one is that it took a good deal of time (more than half a million years); the other is that the neural mechanisms I refer to are present in cousin chimp. Of course, no one knows just when full abstraction evolved; I'm guessing over the period of brain enlargement, but it's just a guess.

In saying that the neural mechanisms are present in cousin chimp I'm referring to the case of Santino the zoo chimp who loved throwing stones at visitors. Lots of captive chimps enjoy throwing things at people, but Santino went further. His keepers of course disapproved of his pelting visitors with stones and cleared his pen of such missiles. But, Santino found others by wading the moat round his dry land feeling for stones underwater. On top of that, he broke pieces off his pen's concrete flooring, which he then broke up into missile-sized pieces. He hoarded these missiles in little piles that were always hidden from the visitors' view. He had to do this after the zoo closed for the day so that there were no visitors present to see what he was up to. He similarly had to do it when the keepers weren't looking, and before he was locked up for the night. In the morning when visitors started coming in he would casually saunter past a stockpile, unobtrusively take a missile and then suddenly hurl it, taking the victim by surprise.

One might explain this behavior (tortuously) as purely learnt; or as an example of abstracting an imaginary future action based on past experience. Fortunately, it doesn't matter for my present purpose, because either way it shows that the neural pathways for planning a future scenario taking into account numerous memories of relevant things, e.g., keepers' disapproval, concealment, time of day, are well established in cousin chimp. There are many other observations and experiments with chimp's and other apes' notorious skill at deception, and I've no doubt that those who work with apes and elephants and dolphins have witnessed many such examples of similar behavior that hint at abstract thinking. My point being that we must have had these all-important neural pathways long before the time of the Big Brain. What the latter did was enhance the potential in the existing neural pathways by either adding *de novo* the capacity for abstract thinking, or greatly expanding a rudimentary capability. What we can't do is identify the precise selection pressures that led to the final product; the archaeological record simply can't single these out. We and cousin chimp passed millions of years without demonstrating an unequivocal capacity for and use of abstract thinking. That we only find clear evidence of it during the Big Braininess doesn't mean that it wasn't there much earlier.

Most scientists who study evolution are wary of what constitutes indisputable evidence of abstract thinking, tending to stick to such prominent phenomena as rock painting. These are seen as reflecting true, fully developed abstraction because they are not an attempt to recreate a memorized scene; there's no landscape behind them. They are independent mental creations condensed from multiple memories of things perceived, as powerful and evocative as anything artists have done since. They are images of things wanted, of mental ideals. Neanderthals had the capacity, but it seems far less developed than in us, if the only known Neanderthal imagery is anything to go by. In Africa, where we did virtually all our evolution, and all of our Big Braininess, we began drawing and painting long ago. Symmetrical cross-hatching scratched onto stones goes back at least 73,000 years. Animal drawings on cave walls go back 44,000 years and by 30,000 Ya we were making skillfully executed paintings of animals, people and fighting. It wasn't until we were well established in Europe that we produced the virtuoso paintings of people and animals in Chauvet Cave 32,000 Ya and more recently at Lascaux 17,000 Ya. One of the things that strikes me as truly remarkable is the perfect perspective in the representation of the two hind legs of one of the aurochs at Lascaux. Perspective has given, and still gives artists generally a hard time, and certainly demands a degree of abstraction. Still, most of the cave images do not incorporate perspective and front or rear views are also rare. Artists from the beginning seemed to prefer to flatten their images, to eliminate rather than render perspective.

There is something about abstract thinking that in my opinion has been overlooked by just about everyone, but which is crucial to understanding why we are as we are today. Because abstract thinking has become part and parcel of everyday life we forget, or, as it is for most people we are unaware of its novelty. We simply do not take account of how recently full abstraction evolved, and of how much of a cloak it is, thrown over our true, anciently evolved selves. I shall pick up this crucial point again.

In cousin chimp societies most males end up being murdered - by their sons, brothers, compatriots or strangers. Death by non-violent natural causes in old age is a rare luxury. Cousin chimp does not complicate matters by feeling guilty about murdering someone, even a relative or someone previously liked. Whatever the motive, it doesn't bother the murderer. Chimp murderers don't anticipate the future by the consequences of the past, i.e., don't perceive that what they just did will probably happen to them. Cruelty, pity, remorse, grief –

these are emotions found only in us because they all involve abstraction, the ability to detach oneself from one's compulsive behavior and relate to others in the same way that we relate to ourselves. Being driven solely by instinct has the enviable consequence of not leaving any mental debris to create such demons as guilt or anxiety.

Evolving as we did an expanded consciousness that included the ability to anticipate the future by reference to the past had a side effect; the knowledge that some unwelcome or frightening past events would inevitably happen again, not only to others, but – much worse - also to oneself. For sure, anticipating the future conferred enormous advantages. That's why it evolved. It also showed us that certain unwelcome events were repetitive or inevitable. Alarming events like dry seasons or winters were manageable by imitating their timetables, thereby convincing ourselves that we were predicting their endings, which dealt with the anxiety they inevitably caused. The devices such as Stonehenge that focused on the exact day of the winter solstice and confirmed that the sun would now reverse its unsettling progress north and put an end to winter are examples of such sympathetic magic. As Frazer so eloquently put it *'Men mistook the order of their ideas for the order of nature, and hence imagined that the control which they have, or seem to have, over their thoughts, permitted them to exercise a corresponding control over things.'* Frazer saw that humans conceived of two sorts of magic. One was sympathetic or imitative magic whereby it was believed that imitating something wished for – an animal, a cure, rain or some such – if done in an appropriately ritualistic way would get it for you (or a paying customer). Individuals set themselves up as practitioners – sorcerers, witchdoctors, healers – thereby creating the first truly professional cadre (predating even prostitution).Sorcerers are still popular in less educated societies and exist even in the heart of the most civilized societies.

The other sort he called contagious magic. People believed that any part of someone's body – hair, blood, nail parings – retained a strong supernatural connection with that person after parting from him. If a sorcerer – or anyone in fact – could get hold of such a body part they could by performing the necessary rituals do anything they wanted to that person via the part. Sorcerers claiming to cast such spells could in actual reality cause individuals to fall ill or even die, so strong was the victims' belief in sorcery. In some societies even one's true name had to be kept secret and pseudonyms used instead as names were an integral part of you.

Contagious magic lives on in such practises as gift giving. A gift given with genuine sentiment (as opposed to traditional presents given mechanically without sentiment) is contagious magic in reverse. The gift is part of you that attaches the receiver to you. One never truly parts with such a gift, which, if not treated by the receiver as indebting him, however slightly, to you, can be deeply resented by the giver.

There was one inevitability that was specially frightening because it was both unmanageable and inexplicable: death. The empathetic awareness of death to kith and kin echoed the realization of one's own eventual death. Death – what every individual instinctively and consciously avoids as the most pressing of all behavior - gradually entered the awareness of self as ultimately unavoidable. Yet it defied explanation. The emotional impact of that awareness was anxiety, fear of the unwanted inevitable and above all the inexplicable. A satisfactory explanation had to evolve if the selective value of projecting into the future was to be sustained.

What evolved was the capacity for belief. Now, belief manifests in several ways and is not one of the clear, hard instincts, like copulation or looking for food. Rather it's one of those instincts that are in the form of a template, a receptive mental function that an individual is born with (genetically determined) but must have the blanks, so to speak, filled in from the cultural background of the day, acquired as the individual develops. In this respect it's like the template for language, or music or ritual. One of the defining characteristics of childhood is the naïve readiness to believe what one is told; the belief function working in its basic form. Gradually, the individual gains self-assurance, knowledge and the capacity to select what to believe. Being social animals we are all of us told by our kith and kin all manner of things that are just so, including the stern admonition to accept that they are just so. Do as you are told. What I say three times is true. Don't ask me why – just do it. Don't do as I do; do as I say. The more primitive or naïve the society the more ritualized and repetitive is this process of cultural transfer, which, it seems reasonable to accept reflects how things were in our past. A great deal of what we are told to believe as we grow up is plainly sensible and easily perceived to be so. We can see for ourselves why we were told to believe that work must be done, that poisonous tubers must be cooked, not to defecate indoors, that snakes are dangerous, that women must not be forced to copulate, that such and such is good or bad, and so forth.

Then there's the class of beliefs that we adopt as virtually unconscious acts of faith in spite of their irrational basis. A visiting alien would be astonished to see us hurtling along narrow roads in opposite directions, specially when he saw how often the inevitable happened. We have faith in such practises, the more so the younger we are, often to the point of delusional invincibility.

Likewise we have faith in surgeons, politicians, judges, living on active faults; in jet airliners, wooden houses, government, homeopathy, Gaia, astrology, herbal remedies and all manner of overtly dangerous or absurdly potty things. It's in our genes, here to stay.

By far the most influential of all beliefs evolved, in my opinion, from the fear of death, the realization that sooner or later we were going to die. One of the strongest – perhaps *the* strongest - instincts is to avoid danger, to keep safe, to stay alive. All animals have it, but only a few, elephants and our ape cousins perhaps, seem at all aware of death itself, that a live animal can die, but remains the same animal. For most animals, a live and a dead individual are two wholly unrelated things, to be reacted to in completely different ways. For us, the comprehension that it was the same individual who was alive, and then dead, was a huge step away from the original simplistic mentality. Specially when that knowledge was reflected onto our own lives. For me, it explains why the ritual of burying the dead evolved, or to be more specific, burying kith and kin, particularly kin. As with everything, the closer it is to you the more evocative. The death of an unknown person who lived in another group had little or no effect; in fact, corpses of strangers or persons you hated were far more likely to be eaten or simply ignored. The closer that person is to you, literally and emotionally, the more evocative the death. Finally, the death of someone dear to you evokes a strong emotion of personal loss: grief. The key point is that of your *personal* loss. As Freud explained it, the sensation of grief arises because one has invested a great deal of emotion in the dead person who was close to you. When that person dies they take with them all that affection and care of yours; you are deprived, robbed, your reciprocal needs and expectations arbitrarily abrogated. The emotional perception of that kind of loss is grief. It's hard to simply abandon that person, to leave them to be visibly torn up and eaten by vultures and jackals. Burying them serves three practical purposes: to protect them, and all that you invested in them, from desecration; to put them out of sight; and to preserve them, to keep them. All of which actions are not taken in the interests of the deceased, but to soften your own unease. You do not like to be reminded of your own eventual death.

Gradually, though, burial became integral to a fourth, much more influential matter. While burying a corpse helped mitigate the living's disquiet – out of sight, out of mind – selection strongly favored another adaptation to coping with the fear of death: the notion that death marked a transition into the supernatural domain. Towards the end of the period of brain enlargement, 130 Kya, burial was a well-established practise. It seems to me entirely reasonable to suppose that the inexplicable process that could, often quite suddenly, transform a live person into a dead one was construed as supernatural, magical. The final step in this mental evolution, the final idealized wish to avoid death, was to conjure the illusion of an afterlife as being part and parcel of the supernatural. If, as I suggest, the supernatural already possessed powerful attributes, extraordinary beings potentially capable of anything people wished to credit them with, it was not such a big step to the delusion of an afterlife. The wish to stay alive, buoyed as it is by the most ancient and forceful instincts of self-preservation is unquestionably a powerful, universal wish, as much today as ever. In the sphere of death and our adaptations to coping with it, for that wish, that ideal to be satisfied, the illusion of the supernatural and its other world, other life, was a simple enough solution. There had to be the template for belief, or delusion waiting to be populated – that had to be already there. Of course, we have no idea when the propensity for belief evolved, but it was a prerequisite for the illusion of life everlasting in the supernatural world.

****

Regardless of the selection factors for belief, in particular belief in an afterlife, it became deeply imbedded in our psyche. Countless millions of hours over untold thousands of years have been spent exchanging a vast kaleidoscope of delusional fantasies of heaven-loads of gods cavorting about the spirit world now torturing us, now coddling us, copulating, cursing, blessing, feasting, murdering, plundering, travelling, fighting, transmogrifying, raging, sleeping, making magic. Whatever people did, or wished to do, to themselves or others, they assumed gods did; only on a bigger, grander, more extreme scale. All the iniquities of Nature as well as its valuables were put down to the volatile tempers of the gods. That of course generated the wish to influence the moods of the gods by sucking up to them with sacrifices and homage. And that was how we ended up with religion.

8.1   Religion

Such are the heights of wickedness to which men are driven by religion.
Lucretius 99-55 BC

Faith is the assurance of things hoped for, the conviction of things not seen
Jesus of Nazareth

Religion. What a strange tale. Has to be told, though, if we are to understand why we behave as we do, for it has been part and parcel of society for a long, long time and moreover isn't going away. Darwin and Wallace gave it a jolt, but not much of one. The people who keep us in their power, our leaders, are still themselves too afraid of god and priests to cut themselves loose and concentrate on life as it really is rather than as priests would have them believe. Before looking at why religion's chokehold on civilization is so difficult to wriggle out of it's necessary to see how contemporary, organized religion came about.

The purest record of the evolution of the wish for immortality into ritualized religion as we know it today comes from pharaonic Egypt beginning about 6500 Ya and extending well into the Christian era. This was more or less the end of the classic Stone Age in Egypt, with farming fully established, but tools still of wood and stone. Though illiterate at the outset, the ancient Egyptians began writing about 5000 Ya, at more or less the same time as in Mesopotamia. From then on religious practise was recorded in various documents including the so-called Pyramid Texts that provide an unbroken history of the evolution of religion extending over nearly 5000 years (half the Holocene). The records were unearthed during the archaeological investigation of thousands of mostly intact graves covering the entire period, all of which consistently preserve certain prescribed burial rituals. Everything revolved around belief in a spirit or soul that passed into the 'underworld' after death of the body, and thus conferred immortality because those who entered the underworld did not die; bodily death was merely a transition. Successful passage into the underworld however was contingent upon observance of certain sacred rituals, principally: a proper burial; and a place at the grave for the living to provide the deceased with food.

These rituals reflected the belief that daily life in the supernatural world was essentially the same as in this world, with the same daily needs plus additional unique obstacles and dangers, and populated by gods and the ghosts of deceased people and animals. The dead person's spirit still needed food, tools and weapons in the afterlife. His grave gave him shelter and a suitable point of departure for the other world. He was therefore interred with all his household goods and chattels and food to be going on with. Above the grave was an alcove with an altar where the living relatives were to regularly provide food and drink. Sometimes, a door was carved in the alcove to indicate that the *ka*, the spirit of the deceased, lived behind it. More elaborate graves had a chamber behind the door containing carvings of the deceased and his family.

After a few centuries actual burial goods were dispensed with by simply carving their likenesses on the chamber walls (after all, why waste sound objects when symbols will do). Soon after that, the onerous custom of proffering food was dealt with in the same symbolic (not to say convenient) way. The Pyramid Texts recounted the supernatural powers of the underworld via the Osiris-Isis and related legends so as to preserve the deceased's body, feed his *ka* and support his counter-spirit – the *ba* - in its dealings with the supernatural. Not surprisingly, the kings - the pharaohs - came to identify themselves via good old sympathetic magic with the gods of the underworld. Assuming divine status before death would ensure a much smoother passage through the tricky underworld and its rather secretive and untrustworthy gods to the afterlife.

During the Middle Empire the power of the pharaohs declined somewhat and feudal baronies sprang up. The barons of course soon reckoned themselves worthy of special treatment in the afterlife, which they achieved by learning the magical formulae, invocations, recitations and mummery that were recorded in the burial inscriptions of the Pyramid Texts. In the New Empire the next inevitable development was that a priesthood set itself up to interpret and control religious practises, especially burial customs. Immortality for the common man, they averred, could only be achieved by initiation (at a price). Naturally, these priests by taking possession of religion ensured for themselves a privileged treatment in the afterlife, as was accorded kings and barons. Finally, burial became a full-blown business with a range of trades retailing mummy caskets, charms, magical texts, body preparation and grave construction.

It's a beautifully straightforward, logical and consistent story with the artful mixture of supernatural belief and everyday human foible so clearly recorded. It's a unique record mainly because for so many thousands of years pharaonic Egypt was a relatively homogeneous, stable society little disturbed by attack, invasion, or insurrection. At the same time, in Mesopotamia, a similar but much less neat and tidy process occurred. Less tidy because by the time they invented their writing 5500 Ya they'd been farming for at least 6000 years - almost certainly much longer in fact. They were way ahead of the Egyptians in this, in consequence of which they'd had rival kings and priests for many millennia, which in turn meant they'd been invading and massacring each other for ages with a concomitant reinvention of society many times over. None of their first writers were interested in this history, nor did they formally record their religious practises, only making what archaeologists refer to as literary references. They recognized at least 2500 gods, but kings tended to appoint a national boss-god – a sort of chairman of the board. Like the Egyptians they had an underworld to which one's soul would go after death provided one was properly buried and kept in food, etc. They saw it as a rather dark and gloomy place – not a great place in which to spend the forever. The really interesting thing is that they referred obliquely to heaven, a great place, but out of reach to most. This is obviously the beginning of the subsequent relegation by the people of ancient Israel of the underworld to the status of hell with its antithesis of heaven.

> If god did not exist it would be necessary to invent him.
> *Voltaire*

In fact, little of the classic Egyptian and Mesopotamian religion has ever been fundamentally changed by western civilization. Sometime after literacy, about 3000 years ago as legend has it, the prophet Abraham left Mesopotamia for Canaan where, seeking a more satisfactory form of deity he consolidated the multiple gods of the afterlife into a single, paternalistic and more predictable god – a perhaps inevitable mental process of condensation as is seen in dreams and memory processing, plus the inevitable wishful thinking. This god of Abraham gave rise to the first monotheistic religion, Judaism, about 2600 years ago. The Jewish prophet Jesus founded Christianity some 2000 years ago and Islam was founded 1400 years ago. The steady symbolization of the burial rituals begun by the Egyptians was continued by the followers of Abraham. At the same time, the 'laws' promulgated by the Mesopotamian kings were turned into lists of observances – commandments. Around 3000 years ago Moses claimed that god had personally given him a much shorter (condensed) list of commandments, which, if adopted and observed by the Jews would ensure them life everlasting. This covenant, contract, between Moses and god was inscribed on stone tablets stored in a wooden box – the Ark of the Covenant. At first the Ark was kept in a tabernacle (tent) but the Jewish King David's son, King Solomon, eventually built the first Jewish temple in

Jerusalem to house the Ark of the Covenant. Or so goes the long-standing legend.

Two thousand years ago the Jewish prophet Jesus of Nazareth renounced what he saw as the remaining barbarities of Judaism and transmogrified the perception of divine belief, rather extravagantly, into a loving god and wonderful everlasting life in heaven (for the faithful that is; infidels were consigned to everlasting hell). He also preached that belief in god was something every individual should accomplish in his mind without the aid of material symbols or devices such as temples and priests. As soon as he was dead though Jesus' refined ideal was corrupted by the self-appointment of priests and their churches that appropriated the role of god's enforcers, just as the Egyptian and Mesopotamian priests had before them.

Finally, 1400 years ago the prophet Mohammed promulgated Islam (meaning submission to god). Islam traces its origin to Ishmael, Abraham's elder son, who is viewed as one in a chain of prophets from Adam to Mohammed. He had seen how the purity of Jesus' interpretation of god had been so grossly corrupted by priests and accordingly forbade any such institution in Islam, including churches[15]. Nevertheless, Islam has become an organized religion like Christianity with all the defects that go with that, even though there's no 'official' priesthood.

The determination of individuals to invent themselves as priests and masquerade as god's proxies is just another way of taking control over people. As Voltaire and others have observed '*The institution of religion exists only to keep mankind in order.*' And bossing people around as an end in itself was an inevitable consequence of surplus – a way of diverting control of some of that surplus to yourself to be disposed of as it suited you. The various factions of Christianity for example have amassed a wealth of some $500 billion – notwithstanding Jesus' caustic denouncement of the worship of mammon and material goods. Such venality has nothing to do with belief in god, though it's of course a vital ingredient of wielding power. Jesus I'm sure would be turning in his grave, were he still in it.

The evolution of present-day organized religion from the primitive belief in the supernatural and an afterlife has taken place in recorded history and as such is there for all to see. It's an altogether lamentable story of the gradual synthesis of a loving god and a flawlessly happy life in a paradisiacal heaven from a risky passage into a dangerous supernatural world ruled by

---

[15] Mosques are free places of worship, not the property of a priesthood.

cantankerous, unpredictable gods of infinite form. I don't see how anyone can fail to notice how this evolution was driven by plain old wish fulfillment. Why else would the original concept be changed into the current one? Nothing else has changed in the last few thousand years that we apes should deserve such a wonderful time as immortals. Freud was right to conclude that belief in god is infantile and primitive. Research has shown that it correlates inversely with intelligence. Intelligent people are more likely to be agnostic, creative, tolerant, open-minded, empathetic than less intelligent folk, who tend to be more authoritarian, less tolerant, more extreme, more likely to believe than to figure out for themselves.

It brings to mind the oh-so delicious irony of the much-loved biblical injunction to 'put away childish things,' interpreted theologically as to eschew immature thinking and behavior and become mature in your thoughts and understanding[16]. To sign up, so they say, to god the father and life everlasting. Yet there can be nothing more infantile than to believe in the tooth fairy, or the god-fairy, or the forgiveness of sins, and goodness and mercy all the days of thy life. The truly immense power of the primitive drive to escape Nature and enter the Kingdom of God, to be loved, needed and so important as to warrant immortality. The mind boggles at such gross wallowing in self-indulgence.

Belief is one thing. Quite another is the use of it to exercise earthly power. The priestly barons of the church weren't satisfied with simply organizing religion, of running a souls protection racket. They saw

> Prisons are built with stones of law, brothels with bricks of religion
> **William Blake.**

that they could also influence the ultimate power holders – politicians - by the simple expedient of threatening them with damnation if they opposed god's policies. Since politicians were either nominal believers or (rightly) afraid of missing out on votes if they insisted on being infidels the priests had them by the short and curlys from the time of the pharaohs on. One of the ways in which priests exercised their power was by assuming possession of the fundamental laws that post-farming society defined after surplus food expanded the all-important principle of property. What Christians know as the Ten Commandments are a highly condensed version of the first written laws that Mesopotamian kings listed by the hundred. The earliest comprehensive list of laws – idealized statements of intent is a better definition – so far found was drawn up by the Sumerian King Ur Nammu 4000-5000 Ya. Fragments of

---

[16]As Jesus neatly put it by way of example, 'Be as shrewd as a snake, as innocent as a dove.'

an earlier list drawn up by King Urukagina cover similar topics. The latter also made himself head of the priests and it's clear that the tussle over power between secular and religious power-mongers had been underway long before writing. These laws or edicts ran to the hundreds and are remarkably comprehensive, including libel, the corruption of justice, all manner of theft and receiving stolen property, the legal position of female publicans, protection of the poor from exploitation, dealings with slaves, etc. The difference between these laws and what we understand by laws today is that the notion of the rule of law had not yet been mooted; Mesopotamian kings were dictators. Legal disputes were in fact dealt with in court by judges whose decisions were not so much based on the royal edicts, as on society's prevailing views of right and wrong.

A side effect of all this was that leaders used gods as a means of unifying the people they wished to control. Inventing religions that pivoted round certain gods and coercing

> An agnostic is someone with no invisible means of support
> *Bierce*

people into adopting those gods and their alleged commands proved an effective way of unifying large groups. The device of ritual, the repetitive reinforcing of specific obligations and beliefs disciplined the followers, kept them in line. It was only recently that kings managed to impose their authority by promulgating so-called laws in writing, punishing offenders not by divine wrath but by brute force. By this device they partially broke the stranglehold of the priests on this aspect of governance and created the principle of secular rule of law; a good example of idealization substituting for reality. Even today, notwithstanding their cherished constitution that stipulates secular government, Americans are obliged to trust in god and put up with presidents who publically cross themselves, brandish the bible and announce new parliamentary bills prefaced by 'god willing'. As I write this, American priests are blackmailing their pious president by denying him communion if he continues to support abortion. Secular government in the US and elsewhere remains a classic ideal in that its actual achievement is a mirage – visible but unreachable.

Over in the Far East the immortality wish took other rather effete forms. Buddhists for example believe that life is a continuous cycle of rebirths leading to the ultimate ideal of release from the earthly life; Nirvana. Such release can be attained only by those, who, after a succession of existences have succeeded in shedding the desire to be alive, revealed by their increasing purity (as defined by the renouncement of earthly or bodily indulgences). One must learn to destroy the perception of life's difficulties & imperfections until your mind is 'pure.' No souls, no god, only personal Nirvana. The ideal of Nirvana is essentially the same as that of heaven, just more wishy-washy.

I should mention that the view of religion as a device to counter the fear of death differs from Freud's interpretation. He concluded that the belief in god – and he was referring to monotheism – arose out of guilt for the primal murder of the father by the sons. For me, Freud's interpretation can't be right, because it refers to monotheism, which is a recent condensation of ancient, multiple gods representing everything people could think of into a single, now loving and loveable, fatherly god. The original concept preceded fatherly monotheism by at least 130,000 years. The last two or three thousand years adds precious little to our understanding of the origin of the god myth.

At the same time we have to remember that we scientists are not above belief; we all have the same genetic template waiting to be filled in. It's an act of faith to believe the intrinsic truth of our perception of logic and reason, the laws of physics, that everything about us is ultimately explicable in terms of the behavior of atoms and the physical forces acting on them. In fact, scientists are guiltier of belief than they often seem to be aware of. Hubble believed the universe was static long after it had been convincingly shown to be expanding. Newton believed in alchemy. Einstein found it impossible to accept quantum physics as essential to theories of matter and energy. Science today is riddled with contradictory hypotheses each of whose proponents strongly believes theirs to be true, dismissing those they don't like as misinterpretations or unproven or fanciful. Our behavior as I propound it here will be dismissed by many as poppycock, I have no doubt. That I believe it to be a fair approximation of the facts of life is an act of faith, however objective I try to be.

As a device for unifying large groups of people religion has proved to be far and away the best we've thought of. As Hitler (and he would've known) noted *'People are far more likely to buy a big lie than a small one.'* The promise of immortality is the best snake oil that mankind's ever been gulled into and remains so to this day. Like all natural remedies it works as well as the user thinks it does. It wouldn't matter if that were all there was to it. It's when it's used to control people and humbug them into doing what priests tell them to do that the trouble starts. Nature is there for you to dispose of as you please, they say – just help yourself. It's why god put it there. God insists that you not impede conception, or terminate it; doesn't matter if you can't afford all those children (and neither can anyone else). God will provide. Just have faith and everything will come good in the end. Remember, your faith is the only faith and those who falsely adopt other faiths, or who refuse to believe, must be damned to hell. The fear of god, as with all fear, is basically the fear of violence.

The menace in this kind of all-encompassing belief is that it encourages apathetic thinking in relation to real social problems. It celebrates a primitive, infantile attitude – in god we trust. Let us pray and hope that divine providence will sort everything out for us. It would seem to me an NSDT statement that civilization isn't going to advance based on trust in the supernatural. It's our brains brought us to where we are and unless we exploit their potential all those millions of years of evolution will just fizzle out before their time. Ashes to ashes, dust to dust.

A final word about belief. Recently, just the other day in evolutionary terms, some of us began to believe in something quite different to the supernatural; science. It's important to remember that science isn't an independent entity. It's simply an intellectual model, established on the belief template, for deciding what is or isn't true, where truth is something universally demonstrable and repeatable. Before science, anything was true that was believed to be true, regardless of whether or not the evidence for it, if any, could be demonstrated to others. No distinction was made between things that **were** demonstrably so, such as bananas or birds, and things that were **believed** to be so, such as ghosts or gods. Science simply separated the two. The pre-science manner of thinking was, unfortunately, not displaced by science at all, and continues today little changed from when it first dominated our mental lives. People still believe in gods and ghosts and goblins and destiny and providence and all kinds of gobbledygook. Science is treated as an extra, an optional app that many, probably most people prefer not to clutter their minds with. It's far easier to simply believe than to laboriously learn, observe, analyze and draw irrefutable conclusions. Science is hard yakka. Belief is a doddle. Belief is infantile. Science demands responsible, mature thinking.

Science enabled us to comprehend the allegedly supernatural and to replace animism with better (if more mundane) assumptions. That it took us a long time to distinguish between fact and fantasy was, among other things, perhaps, owing to the relative strength of our unconscious; instinctual mental activity exacerbated by lack of information. Making the distinction needs a lot of information that each new generation needs to be taught in addition to any new information it may acquire. Amassing and passing on that information verbally before we had writing and paper was inefficient, because of condensation, revision, forgetting, tampering and Chinese Whispers.

8.2   Rhythm and music

Another remarkable evolution during the time of the Great Braininess was that of rhythmic music, a unique attribute of our species. No other animal has a sense of rhythm or can synchronize with rhythm, not even cousin chimp. Chimps and gorillas can be taught to speak pidgin English (in sign, because they're anatomically incapable of uttering the necessary sounds), but cannot learn or even join in rhythm; they simply lack the necessary gene(s). Some think the complex musical devices some songbirds employ include an element of rhythm, but if so it is not used as a socially binding adaptation. The earliest evidence of rhythm is 40,000 Ya when Neanderthals made fipple flutes (recorders) out of hollow vulture wing bones. Presumably though the adaptation that we know as rhythm evolved long before flutes and the absence of any earlier musical artifacts suggests that it evolved as rhythmic dancing timed by tapping or drumming and no doubt singing – as it's still practised by cultures remaining close to Nature. However it came about, the likeliest reason (to me) for its selection was as a coercive device.

The powerfully repetitive aspect of rhythm is analogous to ritual, which is the main device for enforcing and reinforcing cultural practises. Rhythmic music is pleasurable and triggers strong, mostly pleasant emotions. One of its salient features is the extended period over which musicians and dancers can immerse themselves in rhythmic action; all night is common in simple as well as diverse cultures. No other pleasure is as long lasting. The evolution of this ability to rhythmically dance and sing to the point of total exhaustion with no waning of pleasure is most striking. I suggest that this was an important element in the maintenance of structured, relatively uniform societies. Rhythm was something highly pleasurable that everyone could join without generating social conflicts.

Music releases the neurotransmitter dopamine, which is involved in emotional and social functioning by regulating pain and encouraging sociability. It also releases endorphins, which create euphoria, quell anxiety, ease pain and stabilize the immune system.

Group work in today's simpler societies is often accompanied by rhythmic singing. Contemporary urban workmen insist on music while they work, or travel, or do anything not intrinsically pleasurable. The making and marketing of music is a vast industry that like the food industry is self-perpetuating because the appetite for it is only momentarily satiated.

8.3   Our Conscience

At this point it's necessary to introduce another, crucial, element of the Freudian model of our behavior: our conscience, or, as it's also known, the super-ego[17]. It's a brain function unique to our species (though some see its rudiments in other animals) and is undoubtedly (in my opinion) the most influential aspect of what was evolving during the 550,000 thousand years we spent enlarging our brains. The function of the conscience is to veto the expression of raw instinct where that expression would conflict with the cultural morals of the society of the day. For the most part it does this unconsciously, but some of the same or similar restraints can also be applied consciously, as can the opposite: the secondary over-ruling of an unconscious inhibition by the conscious. The mind is nothing if not complicated!

Everyone knows that we have a conscience[18] that prevents us from acting in contravention of an ethical or moral principle. Or tries to. As we all also know, the urge often overrides the conscience. When it does, we experience the sense of guilt. You might succumb to coveting your neighbor's spouse and even act accordingly. If you have a well-developed conscience you will feel guilty to a greater or lesser degree, depending on your powers of rationalization and self-justification. Guilt is the device by which the conscious operates. Something like the conscious could not have evolved if it lacked the means to enforce its vetoes – an NSDT statement for sure. And that enforcement had to work on the individual whether or not his moral or ethical contravention was known to others, i.e., it had to operate directly on his mind.

Guilt is the emotion generated by doing, whether deliberately or not, something that society forbids, or wants renounced. The more we evolved genes for cultural as opposed to instinctual control of ourselves and society, the stronger the emotion of guilt became when cultural taboos were flouted.

17Freud called the instincts the 'It,' the self the 'I', and the conscience the super-self. His translator for better or worse rendered these the Id, the ego and the super-ego, which is how they are known today. As many have noted Plato advanced the same tri-partite model, based on his considerable and penetrating observations of mankind.
18All normal people, that is. A common pathology is a weak or vestigial conscience, as characterizes criminality.

The evolutionary selection for the sensation of guilt seems to me an obvious adaptation to the overall mental environment of cultural control of instinct. Not every failure by an individual to obey cultural edicts is known to others. And none of his longings to disobey are necessarily perceived by others. Clearly, some internal device to hold an individual responsible had to evolve along with whatever external devices there were to control him. That device is the emotion of guilt, which parallels the social determination of culpability. It's an extremely unpleasant emotion and is an example of how strong is the instinct to avoid pain. Observance of the rule of law is a measure of how sensitive to guilt the individuals in a society are. The sensitivity is expressed in two ways – fear of the guilt that would result from society's condemnation, and more powerfully, fear of the guilt that would arise internally even if society was unaware of a transgression. A strict society is not by itself enough to function smoothly; its rules and taboos must coincide with those perceived as fair or just by its individuals. Unless a transgression generates a sensation of guilt in the individual its mere prohibition will be useless. Fear of punishment is the basic means whereby all societies function; guilt, or rather the fear of guilt, is the punishment that more civilized societies use to keep people in line.

Our gallery of morals may or may not include some that are now instinctive (genetically inherited) as well as those that are cultured in us by our parents, other kin and society. Some prefer not to recognize genetically determined morals, but rather to assume all are culturally established *de novo* in every child. The evidence to date is unable to resolve questions such as these; they await a lot more research on genes and brain function. I'm inclined towards Freud's view that we have already evolved an innate sense of right and wrong, of fairness. If we didn't we wouldn't have the instinct to deceive others when we think we might gain an advantage by it. You can't deceive someone who has no innate perception of being deceived, of being taken advantage of unfairly. Cousin chimp is an ace deceiver, and he is supposed to behave entirely by instinct. For my purpose here though it's splitting hairs. What matters is that our conscience and its enforcer, guilt, is the principal mental device we evolved to manage our pushy instincts.

Whether or not some morals have become instinctive I see the formation of the conscious in each individual by the population of a template just as I see each individual acquiring his group's language. This view takes its cue from the universality of the structure of the conscience; the detailed morals and ethics may vary among populations, but the structure and operation is common to all.

****

This brings me to the end of the extraordinary Big Beautiful Brain episode of our evolution during which we evolved or at least put the final touches to our so-called higher senses. These are the faculties that make us different to cousin chimp and all other animals. What I've described is what I believe to be the most influential of the adaptations that evolved during this remarkable period. The annoying thing is that behavior doesn't fossilize. We know what we've ended up with, but we can only guess at the selective pressures and their timing that resulted in the behavior we can reasonably suppose we displayed at the end of the Big Braininess 35,000 years ago. Or more properly the rudiments of the resulting behavior. No doubt, a good deal changed between then and now in that the hierarchy of rudiments changed as we responded to a radically changing environment.

Remarkable as the Big Beautiful Brain era was it heralded a much shorter but perhaps even more profound era of our evolution, though the one probably had to precede the other. What changed during this time led to the abandonment of the hunting and gathering way of life forever, and laid the groundwork for everything we have done (and are doing) since. What to me constitutes the rape and murder of Mother Nature.

Part 6 The last days of the Pleistocene

35,000Ya-11,700 Ya

Chapter 9 The Rape of Mother Nature
(Or Farming, the nice word for The Rape)

Birth of the Landed Gentry (and the relegation of hunter-gatherers to the *gramadoelas*[19] and inferiority)

Never does nature say one thing and wisdom another
Juvenal

Brain size stabilizes – dogs and us - first bow & arrow in Europe - settlement & grain-eating begins, meat on call, introduction of milk - onset of farming - big buildings, astronomy - displacement of hunter-gatherer, land now individually owned – a foot in both camps for a while - Last Tango in Eden for everyone else.

We come now to the fourth influential phase in our evolution, the 23,000 years between the stabilization of brain size to what it is today and the establishment of widespread farming about 12,000 Ya. Whatever the adaptive advantages of bigger brains were, they had peaked, implying that the accompanying behavioral changes had also reached their apogees. Such scarce archaeological clues so far unearthed strongly suggest that by now, 35,000 Ya, we had evolved most if not all of the cultural attributes we have today. We had an extremely pervasive belief in the supernatural and of the power of the gods we populated it with, as shown archaeologically by our burial of our dead. Burial and the internment of artifacts with the corpse also suggests that ritualized religion was likely underway with belief in an afterlife, since this was later to be recorded as the reason for burial. Society used rhythm and music to strengthen repetitive rituals and ceremonies that had become devices for culturally homogenizing groups of like people. As cultural

---

19 The wilderness, the hinterland, the never-never. (Afrikaans)

practises were not genetically controlled, ensuring that everyone obeyed the rules relied heavily on the device of repetition that built, strengthened and maintained shared (i.e., homogenized) memories. The device of taboo and later of organized religion that it evolved into relied heavily on the use of abstract symbols. Abstraction and its lexicon of symbols evolved in my view as yet another socially unifying and homogenizing device. Language was the tool that was used to teach group members the meaning of their symbols by continuous verbal repetition of origin and other myths. The strength of taboos and religion stemmed in large part from the belief in a supernatural world. Referring ultimate authority to the supernatural domain enabled those who personified and promoted authority to amplify its power, because the supernatural had evolved as the repository of abnormal and extreme powers, a belief already deeply embedded in our psyches.

We were painting animals and people on rock faces and sculpturing stylized, fat women that were symbolic expressions of our two most vividly perceived ideals: abundant food and fecundity. We had enslaved dogs in a mutually agreeable manner that provided regular, accessible meat. In hot places some of us wore leather loincloths that concealed our genitalia. (At least, females were made to; right up to the present uncircumcised men went otherwise naked. The circumcised, that is, those forced by religion to acknowledge their inherent nudity were in consequence obliged to cover their nakedness, to conceal the visible signals of their sexuality). In cold places we made up for our ancient loss of thick skins and body hair by sewing clothes out of the thick, hairy skins of other animals. Our larger brains called for more energy per unit effort of finding food, which we achieved by improving our weapons (e.g. atlatls) and presumably our hunting techniques. We cannibalized rival apes, quite possibly focusing on the remnants of the erectus, Neanderthal and Denisovan people who all danced their Last Tangos during this period. We made hand tools of wood, stone and bone for battering, cutting, digging, stabbing and throwing that we used for hunting, feeding and fighting. Another tool was fire that we used for cooking, heating and clearing vegetation. We had rhythm and music to reinforce the ritualization of cultural practises.

All these practises and customs were made vastly more efficient by our use of the now highly developed tool of language; every aspect of everything both physical and mental could be made clear to every individual by sophisticated verbal transmission.

What we know much less about is our social makeup. Some think that initially we assorted in bands averaging about 50 individuals based on kinship, as does cousin chimp today. Whether or not we had begun to form larger groupings – societies - of more distantly related individuals we just can't say, but by the end of this period such larger groups must have started to form around the early farms that were widespread by the end of the Pleistocene. All we know for sure is that by the time of writing half way through the Holocene large groups of people had formed into kingdoms, the more successful of which became the various contemporary races of Mesopotamia.

Of more consequence than anything else we did in this period was to increase our consumption of grass seeds relative to other foods. No other single practise, not even religion, has had as much influence over the course of our recent evolution as this shift to grain as the main food. Its primary influence was threefold: firstly, it provided far, far more energy per unit of time spent procuring food than any other food type, including meat; secondly, it enabled storable surpluses; and thirdly, it enabled people to stay put instead of forever having to move to find food.

The concentration of energy in grass seeds is considerable. Studies of surviving hunter-gatherers such as the Hadza of Tanzania show that weight for weight grain provides 66% more calories than the average mixed wild plant equivalent – a huge calorific advantage. The energy advantage clearly outweighed the many nutritional disadvantages of grain. If contemporary hunter-gatherers are anything to go by their mixed wild food diet results in far less cardiovascular disease and diabetes – both associated with contemporary high carbohydrate diets compounded by high levels of saturated fatty acids from fatty meat and milk. Hunter-gatherers eat less carbohydrate than farmed-food man and most of their fatty acids are polyunsaturated. Several hundred plant foods are available to them providing a variety of nutrients, fiber and phytochemicals that enhance metabolism and reduce cancer. Lean wild meat contains only a fifth of the fatty acids of contemporary domestic meat. Also it provides a desirable 1:1 ratio of omega 6 to Omega 3; domestic meat is skewed towards Omega 6.Although fat has always been a highly prized food because it yields twice the calories/g of lean meat or carbohydrate it was a scarce food for hunter-gatherers, only becoming abundant in domestic livestock.

As with so many evolutionary changes the immediate adaptation often generates other advantages that inevitably are selected for, but which couldn't have existed before the initial adaptation. In the case of shifting to grain what then became possible, and was selected for, was the accumulation of surplus grain. Surplus. The birth of capital. The birth of symbolic wealth. There is nothing one can imagine that would appeal to the human psyche more than

Timeline 5.   Covers the Upper Paleolithic period after brain size stabilized.

Kya	Event
11,.500	Upper Palaeolithic ends with start of Holocene.
12,700	Fifth & last cold Heinrich event, the Younger Dryas. Temperature drops to near-glacial levels within a decade.
14,500	Levantine Natufis in villages of mud & stone huts, stone granaries, stone scythes, stone pestle & mortars. By 10.5 Ya many permanent, walled villages like Jericho all over Middle East.
14,200	Dog buried beside humans. Dog populations increased tenfold at this time, perhaps owing to 'domestication'
14,700 to 12,700	The Bølling-Allerød interstadial. An abrupt warm, wet interstadial during the final stages of the last glacial – the Older Dryas - ending abruptly 12.7 Kya with onset of the Younger Dryas.
15,000	Mongoloids colonise North America via Bering land -bridge, or from Europe by boat 17,000 Ya
16,500	Fourth (penultimate) cold Heinrich event
17,000	*H. floresiensis* extinct. We now sole surviving species of *Homo*.
17,000	Lascaux cave paintings, including aurochs with strong perspective in hind legs.
21,000	Global sea levels rise for next 14 k years, reaching contemporary level (130 m higher) 7,600 Ya. Earliest known bow & arrow in Europe. Spread rapidly through Eurasia.
21,500	Last glacial maximum before current warm interstadial. First eyed needles
23,000	Third cold Heinrich event. Galilee hut settlement, multiple grain grinding, first known braided & wound string, fish net weights.
25,000	Genetic analysis suggests dog association began (no archaeological support)
27,000	Last record of *H. erectus* (from Asia)
28,000	The 'Venus of Willendorf', one of many stylized, fat female figurines of this period.
29,000	Second cold Heinrich event
32,400	Earliest cave paintings of *H sapiens* in the Chauvet caves. Gravettian culture, paintings, burials, grain grinding. Oats, acorns, millet. Starch grains swollen, gelatinized, Consistently heated before grinding
34,000	Neanderthals extinct in Europe.
35,000	Brain size stabilises. Peak of warm interglacial 35,000 Ya. Bone tools, full language
37,000	First cold Heinrich event
40,000	Earliest figurative art from many sites
40,000	Human world population seldom more than 1 million.
40,000	Neanderthals make bone fipple flute, i.e., music well established. Also associated with first ivory, bone, antler & shell tools.
40,000	Start of Upper Palaeolithic, ends 11,700 Ya with start of Holocene.

the promise of the vast wealth contained in surplus grain; the idealized notion of the sheer pleasure that wealth would provide. Thus was civilization as we know it launched; on a torrent of concentrated wish fulfillment (what for many is defined as greed). As an aside, it's always seemed to me a rather feeble way in which to become the Grand Masters of Everything– to quarrel over grass seeds like sparrows.

The first consequence of surplus was that it enabled people to become sedentary – at least seasonally to begin with – which meant they could accumulate in high densities. High density, sedentary groups based on grain grown by a relatively small number of people meant that non-farming individuals could specialize in tradable, skilled services full-time without having to find or grow their own food. By the end of this period and the beginning of the Holocene permanent towns such as Jericho with populations of more than 1000 came into being all over the Middle East. The foundations of the urban socio-economic world we know today had been permanently laid.

We also had another tool that no other ape has evolved, and which came to be used on a vast scale as we grew ever wealthier: slaves. Exactly when we started keeping human slaves is unknown, but we do know that during this period we acquired slaves in the form of dogs. As I think the importance of this adaptation goes far beyond the story of dogs themselves I will say something more about it. Its importance derives from the sheer novelty of it – dogs were the first animal we classically 'domesticated', as the euphemism has it. Not that we necessarily made the first move, as most people believe. I find it far more plausible that dogs 'domesticated' themselves, effectively offering themselves as slaves in return for the food and security that we inadvertently provided. Their insinuation of themselves into our society could well have been an event (which probably spanned hundreds if not thousands of years, remember) with profound effects on our mentality, quite apart from their mundane utility. Initially they must have been vermin, a pest that scavenged food from the outskirts of camps, as did (and do) hyenas, leopards and many small carnivores. But, dogs had (or acquired) many attributes that suited them for life with us that other carnivores lacked. They were more diurnal than nocturnal, had tireless perseverance and an infinite tolerance of abuse. They can digest starches. Dogs instinctively follow the human pointing gesture – a rare ability in other animals. Dogs show considerable synaptic plasticity, the cellular correlate of learning and memory. There is physiological evidence of selection for the neurochemical agents associated with tameness and emotional processing.

As time went by and we became habituated to their presence we started to consciously exploit them for food and, probably, as hunting aids. As the more aggressive and wilder individuals were culled they came to be accepted by us as a useful part of life, not only for food (even today, tens of millions of dogs are farmed and slaughtered for food in, e.g., China, Nigeria, Korea, Viet Nam), but also as guards and hunting aids. More than anything, at least so it seems to me, the slavery of dogs must have been the human perception that generated the initial mental paradigm that led to the cultural practise of slavery of people and of other wild food animals that became part and parcel of our recent evolution. I'm certain that we owe much more to Man's Best Friend than most of us realize.

Our dealings with dogs echo many of the characteristics that define slavery. Though on the one hand we salute them as our best friend (I suppose owing to their undoubted loyalty), we also perceive the state of dog-hood, for want of a better term, to represent the vulgar, brutish side of life. The word cynic is derived from the ancient Greek for dog-like, and the philosopher Diogenes, who founded Cynicism, was called 'The Dog' because of his disgusting behavior[20]. Expressions such as dirty dog, she's a dog, bitch, cur, mongrel, dog-box, dog's dinner, hang-dog, to dog someone, standing out like dogs' balls, dog-days, work like a dog, gone to the dogs, sick as a dog, a dog's life, dog-eared, dogsbody, he's a dog, lap-dog, underdog are all pejorative and reflect our underlying perception of dogs as inferior and uncouth. While such would be the primary perception of any slave (not forgetting the element of self-justification), it doesn't mean that the relationship between master and slave necessarily always stayed that way. Affectionate and respectful relationships often developed even though the primary superior-subordinate element probably always remained. Such is the case with contemporary affluent society's remarkably strong drive to keep pet (i.e., slave) dogs, where the utilitarian element has become weaker and emotional attachments stronger.

As I already noted, we can't put a date on the origin of human slavery, but I suspect it was another consequence of the transition from foraging to farming and the concomitant change in our perception of ownership. All the adaptations I've mentioned so far had a deeply profound effect on our psychic in that they unfettered the instinct that all animals have to possess the food they find. It's an NSDT statement to say that an animal regards the food it finds as belonging to it, and that giving it up to another individual is usually resisted. Instances of an animal (apart from mothers) willingly giving up its food to another, as opposed to having it forcefully stolen, are comparatively

---

[20] Diogenes, in an effort to shed pretension, slept in a clay barrel, defecated and masturbated publicly & generally eschewed anything that he saw as affectation. Mind you, he wore clothes much of the time – purists always overlook inconveniences.

rare. Cousin chimp will do it and the African wild dog and other canids regurgitate food for weakened pack members, as do vampire bats. It's generally accepted that early on in our evolution we fed our children (after weaning), and each other depending on relative need and food availability. There is a 1.8 million-year-old *erectus* fossil of a toothless man who had apparently survived long after he lost his teeth, which suggests that his companions helped him eat.

This facilitative behavior did not come about as a result of any collapse of the underlying instinct to keep the food you found for yourself. Rather it was one of the inclusive fitness adaptations that evolved in our overall group behavior. The perception was created of the wider value of food and of the advantage to be had from treating food at least in part as a common good.

I said that the changes arising out of the increasing use of grain food freed the primary instinct in every individual to treat the food he found as his; what we label property. This needs some explanation, specially as it marks a deeply profound change in the cultural structure of society. What I mean by fetters are the cultural restraints on food as personal property. Up to 35,000 Ya, we were all hunter-gatherers. All our cultural inhibitions and redirections of instinct acted on the individual in the context of foraging from Nature. Which is to say that food in our perceived world was not made by us. When an individual killed an animal or dug up a tuber, he was not taking these things from another human, or making them. Having found them he was bound instinctively to want to eat them. At the same time, selection had long ago fixed a suite of genes that adapted him to operate in a group, membership of which conferred many advantages on each member. One of these advantages was that if some of the group found more food than they needed for themselves the others could eat it. This led to selection for an individual to actually bring surplus food back to the group, an action that required energy to be expended without any apparent gain to the individual expending it – something that goes against the grain of natural selection generally and of competition in particular. But, the gain was a deferred one; each individual would from time to time benefit from others bringing back food, hence a positive selection for the trait. This is an example of what is known as ***inclusive fitness.***

Socially, though, the adaptation created a problem in that the bringer of surplus food was bound to regard it as his, which could result in fighting over who could eat it. Selection thus favored inhibiting the natural instinct of ownership of food (which, one has to remember, is when living in the wild state the most valuable of all things). That is the basic reason why in all hunter-gatherer societies wholly dependent on foraging there's a powerful cultural restraint on owning surplus food. A man may kill a large animal that he can only eat a small part of. Rather than leaving the rest where it fell he brings it to the group (or vice versa), but cannot assume ownership of it. Neither can any other individual; it must, culturally, become common property. The same conditions applied to leadership of the group. No individual could assume control of the group. Selection favored leadership in the form of group knowledge, group decision, and group control - the so-called, much exalted egalitarian nature of hunter-gatherer society.

Eating grain changed all this as soon as we started intentionally growing it. I don't think there was any sudden eureka moment when the cultivation of grain dawned on the human psyche. The notion that our ancestors consciously hit upon the idea of taming plants and animals and then set about putting it into practise is to me not only highly improbable but also a superfluous hypothesis (and a typical human wish to fit things into pre-conceived ideals). I see it as far more likely because far less genetically demanding that it simply happened. It's simpler to imagine hunter-gatherers living on floodplains or lakesides in a warming climate where grasses were abundant finding themselves repeatedly collecting and eating the seeds of the same plants in the same places. Defending these valuable localities from other hunter-gatherers became a culturally propagated practise that gradually led to conscious actions to improve the growth of preferred plants at the expense of unwanted ones. Competing plants were removed and what was seen to favor growth – watering, weeding, protection and planting - was imitated and then culturally perpetuated. The advantages of such artificial monocultures in more food were enormous and rapidly attained. Once started, it's easy to see that it could, culturally, be quickly (i.e., within just a few thousand years) incorporated into society.

Australian Aborigines illustrate this paradigm perfectly. When Europeans settled like locusts on unsuspecting Australia in the late 1700s its people had been living there completely isolated from the rest of the world for 50,000 years or more; i.e., since the arrival of our Denisovan cousins in Southeast Asia. Once the sea level started rising again 21,000 Ya and cut Australia off from the north the Aborigines couldn't have known what was going on in Mesopotamia and the Nile valley. Yet the Europeans found them living in many places in villages[21] of 1,000 people or more. These dwellings were similar to those of

---

[21]To the invading Europeans' horror. These villages were quickly destroyed and references to settled or organized existence suppressed in order to preserve the rationalization of *terra nullius*.

present-day rural Africa with walls of stone, logs or branches, and roofs of branches, all plastered with mud. There were doors, chimneys and granaries. Some of these settlements may have been permanent, such as the fishing village at Lake Condah; most were probably seasonal. It's an NSDT statement that such settlements had to have a permanent or seasonal food supply to supplement the foraging the people still practised, some clans exclusively. These seasonal villages were based on grass seed and tuber harvesting, sometimes practised on a large scale. It's not clear whether the various grass grains were consciously farmed or simply natural, though there's some evidence of woody vegetation clearing. What is documented is that mature grasses were cut (with stone scythes), stacked, and subsequently, when dry, threshed and the grain carried to the villages and stored in granaries. Repeated digging for tubers with sticks had resulted in extensive fields of virtual tuber monocultures as other vegetation was progressively removed and the soil in effect hand-ploughed.

The reason why cultivating grains (and other plants such as fruits) caused a regression to a more primitive concept of ownership was the shift to staying put rather than constantly roaming. The individual began to expend considerable effort growing food, and no doubt, defending it from not only wild animals but other humans. In other words, food was no longer just a randomly found part of the common environment, but a fabricated artifact. Sharing it beyond an individual's extended family no longer conferred the old advantages because every individual now had the same opportunity to grow their own food. The grower no longer needed to rely on others and naturally projected that attitude onto others. The fact that it was easy to grow surpluses made everything more secure and predictable. The surplus took on a value of its own in that it could be used to barter for goods and services the cultivator hadn't time to procure himself. It has to be borne in mind that farming wasn't any easier or less time-consuming than foraging – almost certainly quite the opposite. But, it was infinitely more rewarding. The use to which surpluses could be put enabled growers to support people without land in return for their subjugation to the grower. These people could live in settled groups– villages, towns - and provide labor and artisanal services. Surplus food meant that these groups could become much larger than the old foraging groups and therefore better able to defend themselves from competitors.

The evolution of the landed gentry was well on its way. The more ambitious barons of course became the first kings. The individual's perception of the advantages of a farming-dependent society and the generation of large, settled societies clearly outweighed any disadvantages, as it still does today. It was, quite literally, an earth-shattering transition, because it not only heralded the onset of what we insist on calling civilization (from the standpoint of the landed gentry and their acolytes) but also the shattering of the earth's surface (from the standpoint of those of us who fear the consequences of 'civilization', plus most other animals). For some the Rape of Old Mother Nature was simply her Manifest Destiny; for others an unconscionable crime. But, I digress.

One immediate knock-on effect of farming was that those who either had to continue the foraging life, or preferred to, were forced off the farmable land and consigned forever to the *gramadoelas*. As such, they were automatically regarded as inferior in that they retained the barbarous ways that farmers had elevated themselves out of. Jumping ahead a bit, I quote from the first known written description of hunter-gatherers. It was a Mesopotamian king 4400 Ya who wrote:

'The Martu who know no grain... who know no house nor town, the boors of the mountains.... The Martu who digs up truffles... who does not bend his knees (to cultivate the land), who eats raw meat, who has no house during his lifetime, who is not buried after death.'

That one of the first written records ever made should be a civilized king's racist diatribe has a good deal to say about the character of our civilization.

The earliest archaeological signs of full-time farming are around 12,000 years ago in the legendary Fertile Crescent, a region including modern-day Iraq, Jordan, Syria, Israel, Palestine, southeastern Turkey and western Iran. Multiple, genetically distinct populations of people in the Fertile Crescent started farming – not a single population. These didn't interbreed much for a few thousand years (they probably concentrated more on fighting). The first farmed plants were wild barley, peas and lentils, and the first slave animals after dogs were goats and aurochs (wild cattle). Some thousands of years later farming spread into surrounding regions, including Anatolia and Egypt, Europe, southern Asia, and parts of Arabia and North Africa. It's interesting that the spread was by the dispersal of the original farmers into these regions. It was a long time before the locals adopted the new arrivals' farming practises.

Once full-time grain farming became widespread in the Fertile Crescent a suite of animals was progressively enslaved (domesticated, as we softly say) as well. From about 10,000 Ya aurochs (cattle as we now know them), then sheep, goats and boar (pigs) were grown in captivity. Just how we went about this isn't known, but the crucial stimulus to restrain them seems to me to have been our becoming sedentary. Once we were stuck with our grain fields having to chase wild animals for meat became increasingly difficult, because it required time and labor that was no longer available. Hunting became one of the many specialized occupations that were a spin-off of surplus and settlement. To confine free-ranging wild animals and control their movements was an obvious solution. As more and more land was occupied and controlled less and less was available for wild animals so that the conversion of wild to captive animals was itself a function of the grain-driven expansion of human populations. We already had slave dogs that provided a mental perception of the value of captives as a more cost-effective way of procuring meat. At around the same time the making of pottery containers became widespread which enabled the use of milk – a novel and efficient food that had probably never been used before (after weaning, that is). Prior to pottery, there's no archaeological evidence that we had containers for liquids other than things like ostrich eggshells, tortoise shells and possibly gourds. Remnants of the latter only go back 10,000 years or so, but it seems extraordinary that we hadn't used them for much longer. We may have made containers from animal skins – it's hard to imagine we didn't – but if we did none fossilized.

Exactly how we forced wild herbivores to do as we wanted is unknown, but I see no reason to suppose it involved anything remarkable. In Australia, cattle, sheep and goats readily go feral and live entirely as wild animals. The only difference between them and their domestic counterparts is that in the wild they avoid people; as domestics they are tame and relatively easily controlled. My guess is that the Levantine farmers began by enclosing small numbers in *bomas*. Once tame they could be herded by people (small boys, probably) and taken to pasture and water. At day's end they would be herded into a dedicated *boma*, or more likely a common *boma* with the humans.

There was one other thing that began as the Stone Age was giving way to the Holocene; kings started making enormous structures unrelated to everyday living. The oldest known is the symmetrically arranged complex of pillars at Göbekle Tepe in Turkey that is 11,500 years old. There are more than 200 pillars in about 20 circles. Each pillar is up to 6 m tall and weighs up to 10 tons. They are fitted into sockets hewn out of the bedrock. Later pillars are smaller and stand in rectangular rooms with floors of polished limestone. Although no signs of ancient farming survive at or near the site it's an NSDT statement that the enormous effort of quarrying, cutting and dressing, transporting and erecting the huge blocks could not have been made by people whose waking hours were spent finding food; the labor for Göbekle Tepe must have been fed. The only apparent way of doing so would've been farmed grain surpluses (interestingly, the variety of wild wheat ancestral to modern wheat comes from nearby).

We do not know why such vast resources should've been diverted into building Göbekle Tepe but the fact that similar structures became more and more common and were eventually identified as temples suggests that it served as a venue for transactions with the supernatural. If so, it's a testament to the enormous importance of our belief in a supernatural world.

It's an odd thing that the Big Beautiful Brain we proudly sport evolved such a long time ago, because we intuitively associate it with cleverness, which itself seems to us to have increased enormously since we began to farm. This is an illusory perception. Full-time farming was 25,000 years after brain size stabilized – a long time. If other brain parameters were selected for during that time we have no idea what they were. It was farmed surpluses that led to settled life and urbanization that in turn created the conditions for 'cleverness', i.e., made the 'surplus' time that large numbers of relatively crowded people exploited to do 'clever' things, turbo-charged I have no doubt by the effects of crowdedness on each other. And that led to 'civilization', as it pleases us to call it.

Conventional history is misleading because all the recent things we do that seem to us so remarkable are obviously not in themselves drivers of big brain selection; in other words they're not so much qualitatively as quantitatively remarkable. The qualitative changes to our brains evolved over millions of years and little has changed for tens of thousands of years. It's simply that quantity is proportional to the time available for its production, itself a function of the wealth that began to accumulate after the farming of plants. Only the wealthy have more time than is needed for subsistence. We had long-since manifested the core of our intellectual apparatus – logic, reason, belief, superstition and gods, morality, creativity. Writing enabled the recording of history, but the first writings were already sophisticated, i.e., writing was the novelty, not the material written. All that had been gone over verbally long before, at least since our brain stopped enlarging 35,000 Ya; probably long before that. We just don't have any relics to go by; thoughts don't fossilize.

****

So here we were, the hunter-gatherer life gone forever, at the start of a completely new life that would bring us to where we are today. Farming as the primary occupation would gradually give way to an increasingly urban existence that while it depended on farming for its food would find a vast array of other activities made possible by farmed surpluses, the basis of wealth. The only thing that stood in the way was Nature.

Down the highway
Down the track
Down the road to ecstasy
We couldn't help it
Twas in our genes

My apologies to Dylan

Part 7 The Holocene

11,700 Ya -the present

Chapter 10   The Murder

Want, want, want, take, take, take, more, more, more

He that loveth silver shall not be satisfied with silver; nor he that loveth abundance with increase: this also is vanity.
Ecclesiastes 5:10

Holocene begins with end of Ice Age – hunting and gathering abandoned for grain farming - Mother Nature's murder to make way for farming – Noah's Great Rainbow, biblical version of murder – today, half all habitable land farmed – exponential population growth and destruction of Nature - 10 millennia to go from two million to 7 billion – all remaining rain forest will be gone in 80 years – another 3.2 billion people over same period - change in ownership – leadership now by individuals – slavery – unification, coercion and the myth of cooperation – religion and coercion – language and literacy.

With the extinction of our remaining two-legged relatives, tough, enterprising *erectus*, burly Neanderthal, shadowy Denisovans and the gnomish Floresian dwarfs, we had arrived at the end of a long and bumpy evolutionary road – the end of the last Great Ice Age and the end of life in the Pleistocene as itinerant opportunists, living on what we could garner from Nature. We were about to begin resisting the random push and shove of domineering natural selection and choose instead to force things to go in directions that we in our new-found wisdom figured would vastly increase pleasure and decrease pain. We began in other words to become 'civilized' as we nowadays imagine ourselves to be, a process that included an extraordinarily ferocious and relentless attack on the natural world that had been our home and sustenance for so long. A monumental act of treachery, as I cannot help but see it. Or is it just an extreme version of the age-old rejection by the new generation of their fuddy-duddy parents' traditions and values. Or is it just evolution, nothing more, nothing less. Or a tyro god perhaps, stuffing up his first shot at a creation.

Actually it bears an uncanny resemblance to what we usually did when we overran the enemy and sacked a city. We destroyed it rather than simply take it over. Specially the temples. We always knocked down the temples and built ours over the ruins, just as each passing dog pisses on his predecessor's pissed-on pole. It's as if we possessed ourselves of the illusion that the ancient, primitive Temple of Nature to which we had always humbly submitted was now ours to destroy and replace with our own Temple of Infinite Prosperity.

Mother Nature didn't care for as an impartial parent she didn't favor any of her children over the others. If some of her offspring chose to behave like idiots that was their problem, and there was in any case no going back. Evolution's gearbox lacks reverse. Magnificent reptiles had presided over the world for millions of years, yet suffered a truly Shakespearian dusty death just the same. Other organisms took up their atoms and leggoed them into new self-propelled robots of Nature, as will be done with our atoms when we finally fly too close to the sun.

As soon as we realized that we could make the wild plants we had been eating for millennia grow where we wanted, and by destroying all other plants grow in much greater quantities, we could accumulate surpluses. That in turn allowed us to do two things in particular that are, for me, the principal elements of civilization: stay put, instead of forever chasing after food; and force other individuals to make surpluses for you. As soon as you had a surplus of food you had something you could use to your advantage: in effect, money. The advantages (i.e., pleasures) of settlements obviously outweighed any disadvantages since virtually all people given the choice prefer urban to rural living[22]. But, it was another door that opened that may not be quite so obvious: the control of people. The first problem that grain farmers ran up against, apart from the intricacies of husbandry, was labor. Not just farm labor, but, in my view far more importantly, labor for defending the land on which the crops grew. Ultimately it's the land that generates the surplus, then as now. I will return to this a little later.

Having reviewed our history and how we behaved up to the time of full-time farming we can now look at what we did next. I said early on that with farming our species broke with Nature's conventional competition and set off on a course profoundly different to that of all other species. What we did – and are still doing – was to seriously dismantle Nature, our home and provider since we had scampered about as diminutive australopiths all those millions of years ago. Nature suddenly became the enemy, to be opposed, neutralized

---

[22]In 'developed' countries 95% of the population is urban

Timeline 6. Notable events of the Holocene.

Ya	Event
Tomorrow	Rest of our forest cleared by 2200. Ready now to violate another planet. Let's start with Mars
Today	Over half of all land cleared, along with many species of animals & plants. Like many rapes, ends with murder. Population 7.7 billion & counting. Around half are on brink of poverty, 1.3 billion in extreme poverty. 30% of potential labour force unemployed.
120	Internal combustion engines applied to wide-range of Nature-busting tools
154	Dynamite invented. Now we could really blow Nature to smithereens
220	Human population passes 1 billion
1,000	Gunpowder invented (China). Now we could kill people & other animals in even greater numbers.
1,400	Mohammed founds Islam based on Mosaic monotheism, i.e., back to Old Testament ways
2,000	Jesus splits from Judaism to found Christian monotheism
2,509	Roman Republic founded after overthrow of Roman Kingdom.
2,600	Greek civilisation peaks: Laws, democracy, mathematics, poetry, drama, mythology
2,700	Iron age starts in Europe
2,950	First monotheism. First Jewish temple (Jerusalem). Later replaced by Al Aksa mosque.
3,000	Human population passes 50 million
3,500	First alphabet (North Semitic, Palestine & Syria)
3,800	Babylonian civilisation in Mesopotamia. Shang dynasty in China,
3,800	First records of contraception. Great idea - shame it didn't catch on
4,000	Start of bronze age in Europe, first metallic money
4200	200-year period of worldwide cooling & drought
4,700	First Egyptian pyramids. The Great Pyramid of Cheops still the biggest building ever made.
4,800	First writing on paper – papyrus in Egypt
5,000	First written words: hieroglyphics (Egypt), Cuneiform (Mesopotamia). Stonehenge built
5,100	Beginning of Old Kingdom (Egypt), urban Sumer civilisation (Mesopotamia). Irrigation well established in Egypt. Still only copper, no bronze
5,300	Official end of Stone Age with start of Bronze Age
5,400	First use of decimal number system (Egypt)
5,500	First use of wheel (Mesopotamia), for transport & pottery
6,200	Sudden decrease in global temp. Temp had just been 1 degree higher than today. First use of wool
6,350	Domestication of horses (Ukraine)

7,000	Human population passes 5 million. Oldest astronomical stone circle at Nabta Playa, Egypt
7,500	Cultivation of millet & rice (China). Megalith structures in France
8,000	Spinning & weaving (Mesopotamia) flax textiles. First irrigation in Mesopotamia. Cats thought to be associating with humans.
8,200	First copper smelting.
9,500	First walled town: Jericho, Palestine, population 2500. Long-distance trade in obsidian, i.e., persistence of Stone Age practises.
10,000	Continental ice-sheets melt in Europe & North America
11,000 10,000	'Domestication' in Mesopotamia of aurochs, boar, sheep, & goats. 1000 y later of goats in Iran & pigs in Thailand. (Pigs done twice independently) Pottery ovens in Near East. First known cultivation of wheat & barley, in Mesopotamia
10,300 Also 10,000– 9950	Holocene starts with rising temp & rain, rising sea levels. Rate of glacier retreat & rising temp higher than past century. During deglaciation, the equilibrium line altitude ascended 1430 m & the air temperature rose by 9 °C. Deglaciation occurred in two phases. During the second, faster phase, which lasted 500 y, the glacier length decreased at an average rate of 1.7km/100 y, implying a warming rate of 1.44 °C/100 y, indicating a rapid climate shift.
11,500	First large stone monuments, at Göbekli Tepe, Turkey.
12,000	'Official' adoption of farming as primary human food base. Population 4 million
12,000	Fifth (last) cold Heinrich event. Sorghum & millet stored at Nabta Playa, Egypt
12,700	Younger Dryas, a cold period that knocked temperatures back to near-glacial levels within a decade
14,500.	Older Dryas Bølling oscillation. A cold period in the middle of the Bølling-Allerød interstadial
14,700 to c. 12,700	The Bølling-Allerød interstadial. An abrupt warm, wet interstadial during the final stages of the last glacial

and ultimately destroyed. The attempt to murder helpless Old Mother Nature began. No other animal modifies its environment by removing all living organisms not completely enslaved to it. Many animals and plants modify their environment to suit themselves, but not by extermination. Beavers and elephants, for example, can force their immediate environment to cycle, as can other animals. But, they move on (rather they *used* to move on) and it recovers, because they do not stay put until everything is dead. We, though, want *all* its wild animals and plants gone because we want the land they lived on; we shan't be moving on. All signs are that the attempt to murder Mother Nature will inevitably succeed.

The Holocene is the period from 11,700 Ya, when farmed grain replaced foraging as the primary human food base, to the present day. In ***just ten*** *millennia* our population on grain ***doubled twelve times*** from two million to seven billion. It took us our entire evolution less a couple of millennia to reach our first 90 million – now we add that many people ***every year.*** Which gives every Great Helmsman on the planet the contented, warm, furry feeling that comes with boundless surplus. Some of them now feel wealthy[23] enough to invade other planets to plunder yet more good stuff. With god's blessing, of course – Noah's Great Rainbow spans all of Nature, not just this planet's.

The prophets of monotheism were quite lyrical in their accounts of the early relationship between men and god. The story of the prophet Noah's Great Rainbow relates god's disgust at mankind's licentious behavior and his decision to wipe out everything with a massive flood. Because Noah was a good and pious man god warned him of the flood and told him to make an ark and stock it with representative animals. After the flood, god made a covenant with Noah.

> *'Then God said to Noah and to his sons with him: "I now establish my covenant with you and with your descendants after you and with every living creature that was with you—the birds, the livestock and all the wild animals, all those that came out of the ark with you—every living creature on earth. I establish my covenant with you: Never again will all life be destroyed by the waters of a flood; never again will there be a flood to destroy the earth."'*
> *'Then God blessed Noah and his sons, saying to them, "Be fruitful and increase in number and fill the earth. The fear and dread of you will fall on all the beasts of the earth, and on all the birds in the sky, on every creature that moves along the ground, and on all the fish in the sea; they are given into your hands. Everything that lives and moves about will be food for you. Just as I gave you the green plants, I now give you everything."'*
> *'And God said, "This is the sign of the covenant I am making between me and you and every living creature with you, a covenant for all generations to come: I have set my rainbow in the clouds, and it will be the sign of the covenant between me and the earth."'*

For me, there are intriguing conclusions to be drawn from this myth. It states for example that Noah was a farmer, the inventor in fact of vineyards and wine. It refers, I think, to the transition to farming and sanctifies the destruction of nature to make way for it. It reflects unease at the widespread destruction of Nature that accompanied farming, a harking back to the former respect for Nature. The myth rationalizes this anxiety by sanctifying it with

---

[23] All Great Helmsmen regard the money got by taxing the people as theirs to spend as they want.

Noah's Great Rainbow – a move that has worked extremely well, it must be said.

As an aside, it also alludes to a consequence of the invention of booze in that Noah was in the habit of coming home legless – a weakness that was to become one of the defining features of our civilization. (His drinking also led to a regrettable incident with his son, Ham, that I won't go into here.)

It's hard (at least I find it so) to picture the state of the world's habitable land by scanning numbers in a timeline (I've summarized these in Timeline 7). Thinking of it as the murder of Mother Nature it represents a simple crime scene with only a few key features to the motive and MO.

Timeline 7.		The current status of the world's land.	
AREA Millions km^2	%	USE	COMMENT
Earth's surface			
149	29	Is land	
361	71	Is sea	
Of 149 million km^2 total land			
104	70	Is habitable.	
45	30	Is uninhabitable	Desert & mountain top
Of the 104 million km^2 of habitable land			
51	49	Is already farmed	
39	38	Is forest (partly inhabited)	7 are protected (at least for now). 13 are pristine. Approx. 50% of original already gone. Remainder gone in 80 years
12	12	Is dry savanna (partly used)	
1	1	Is built up	
1	1	Is lakes, rivers & dams	
Of the 51 million km^2 of farmed land			
39	77	Is farmed for livestock	Yields 17% of human calories, 37% of protein
12	23	Is farmed for grain & other crops	Yields 83% of human calories, 63% of protein

The world has 104 million km^2 of habitable land – this is the crime scene I'm speaking of. Half of this is already farmed. Of the remaining half, 14% can't be farmed because it's taken up by towns, roads, lakes, rivers and dams (2%), or too dry (12%) for commercial farming (though it's used as much as possible by subsistence people).

The rest is forest – about 38% of all habitable land. That's 39 million km^2, which is quite a lot of land. Bear in mind, though, it's **only half** of what was there a few thousand years ago when we first rolled up our sleeves, sharpened our axes and set to chopping. It was frustratingly slow at first – hand axes have served us well for a long time, but forest trees are big things. Then, a century ago, we perfected the internal combustion engine. From then on the trees really started coming down at a rate that will see the last of the rain forest topple in about 80 years' time – the end of this century. The 3.2 billion extra people we'll have at that point, on top of the 7.7 billion we have now, will need that land to farm the food and create the surpluses that our politicians need to drive the open-ended economic expansion they are mesmerized by.

The destruction of the natural environment was not an end in itself to begin with. It was a consequence of farming, something that was bound to happen once farming became universally *de rigueur*. It was a perfectly normal thing for a farmer to do – and remains so to the present day. Nature for a farmer is a nuisance at best and an enemy at worst. The source of all weeds, vermin and disease. Above all, a barrier to more – more land, more surplus, more wealth. Whether or not one judges our malevolence towards Nature good or bad one must accept that it arose quite simply and quite naturally for a perfectly sound evolutionary reason. As hunter-gatherers we were an integral part of Nature. Nature was the source of **all** food and materials - everything. As such, we loved Nature and sought to understand, anticipate and honor it, to attribute emotions and purposes, to placate it when cross, to celebrate its largesse. Nature was **needed** by us. Above all we saw it as superior, with ourselves in awe, needing to behave humbly and respectfully if we were to get what we wanted from it. In a word, we were obsequious to Nature.

Then suddenly (by suddenly I mean five, ten, twenty thousand years – maybe more) we partook of the forbidden Tree of Knowledge and got ourselves booted out of the Garden of Eden where Mother Nature had for so long watched us gambol around. Her lambs now were headed for the slaughter, to the abattoir of innocence, where all naiveté was stripped off, replaced by self-importance, lust and greed, the lust for more, more, more, always more, never enough. The taboo knowledge, the all-corrupting sin, was the knowledge that we could murder Mother Nature and take possession of our own destinies. We could do better without the old cow. We knew now what we really wanted, and would not let Nature and childish dependency hold us back any longer.

The Australian Aborigines' part-time use of grain well illustrates this scenario. They had not started destroying Nature to make way for farming at the time of the white man's invasion and occupation of their land. They were on the brink of it, but still with a way to go. Love of land, of all things natural still dominated their weltanschauung. They were still in awe of Mother Nature. That is why to this day they find the whitefella's cavalier attitude towards Nature, his contempt for natural landscapes utterly repugnant, even disgusting. Because they retain the ancient hunter-gatherer's love of and affinity with the land and everything on it. That is why *terra nullius* was and is so repellant and incomprehensible. How could we not see ....

One of the inevitable consequences of surplus food was that more children survived to breed, because surplus food eliminated starvation and settlement made life easier and reduced mortality indirectly arising out of constant movement, lack of shelter and predation. It's a universal characteristic of Nature that animal populations boom and bust rather than stay constant. All animals must have the potential to produce more offspring than they need when things are going well. It's a must, because when conditions are bad population numbers decline. When they improve again, those lost must be replaced.

All species of sexually reproducing animals with equal numbers of males and females (us for example) only need two offspring per female to maintain a population indefinitely – provided that those two survive to breed. In other words, if there's virtually no early mortality – a rare circumstance. Even during boom times there's always some mortality, though it can be quite little. At the worst of times, mortality can be catastrophic. That's why all animals must have the potential to produce more than two surviving offspring per female. In our case, we can readily produce ten (or more) offspring per female – five times the basic two. In the case of many fish and many other animals each female can produce millions more than the requisite two. That potential means they can boom again after a massive bust, though both decreases and increases can be quite modest rather than spectacular booms and busts. Grain, slave animals (and people) plus the clearing away of Nature triggered the mother of all booms for us, a boom we are still responding to in the classic, genetically determined zoological manner by producing as many offspring as we can. Of course, for all other animals still dependent on Nature it triggered the exact opposite; the mother of all busts.

It's estimated that over our entire evolution around 100 billion people have lived and died, and nearly 8 billion more are currently alive. Until the peak of the last glaciation 21,500 Ya it's estimated that there were rarely as many as a million people in the entire world at any one time, and sometimes fewer than 100,000. Given our reproductive potential (as foragers) of around six children per woman the fact that for millions of years we hovered around just two surviving offspring per woman underlines how precarious the foraging life was. Starvation, fighting, predation, accident and disease put paid to most people. Mortality was almost certainly highest in early life, just as it is today in cousin chimp bands, which experience comparable death rates.

Our population increased exponentially from the beginning of the Holocene (the Horrorcene) to the 1960s at a dramatic rate, after which it slowed a little. That's because in countries where the population had become urban, women spontaneously curtailed reproduction to the two per female or even less level. Women in the 'civilized' town are more emancipated and less obliged to reproduce for many reasons that are common knowledge. This reduction in breeding was spontaneous and had nothing to do with the leaders whose lust for more surplus hadn't in any way abated. As the global population was still growing fast a large part of the wealth they pursued was generated by that growth. The more they expanded the more they consumed and produced, and the more they suited politicians of all persuasions.

These changing circumstances will affect the future global population in ways that we can only muse over. The more assertive people become the more land they will want. The leaders, to quell the assertiveness of the led, will want yet more surplus to create ever-expanding, more appealingly glittery economies to keep them satiated. The remaining rain forest is on good land with high rainfall – just what's needed for more farm surplus. Economic and population growth is not going to stop of its own accord, or because a minority want it to. On the contrary, all leaders will do everything they can to perpetuate it. The remaining rain forest could last a little longer than 80 years – or a little less. Either way, civilization as we know it must get rid of it. Human growth is sacred; Nature is profane.

And with the clearance of Nature's vegetation of course go all the other animals we grew up with as we evolved, for whom we now feel nothing, notwithstanding our long acquaintance. We no longer feel for them because we no longer need them. We have already forced many large animal species, and half of all plants into extinction by, literally, pulling their carpets from under their feet. The remainder is moving steadily down Desolation Row, keeping time with the rhythmic thud, thud, thud of toppling trees.

10.1   Ownership, Leadership and Slavery

The shouldering aside of the hunter-gatherer by the farmer that had been building up for the preceding ten or twenty thousand years marked a radical change in our relationship with Nature. The essence of this change was a reversal of roles. Up to this point Nature had dominated us and we were the inferior party, utterly dependent on its goodwill. Now the shoe was on the other foot. We were in charge and Nature was the inferior one, to be bullied and beaten as inferiors always are by their superiors. What we now wanted was Nature's land - not the things on the land that we had wanted as supplicants. This didn't change our underlying instincts in any way, but it radically changed how we manifested those instincts. It changed because land is not a consumable good, as all the things that grow on it are. Land – what came to be known by civilization as real estate – is the ultimate possession, the real source of wealth. As farmers we soon came by that understanding. Not only is land the determinant of wealth it also has the far-reaching characteristic of being immovable. These perceptions changed our behavior in relation to the basic instinct to keep what we found for our own consumption, because what we now 'found' – a piece of land on which to grow grain – was permanent, fixed, indivisible and of itself inedible. The ancient foraging perception of the land as something all had freedom of access to was no longer tenable. The tragedy of the commons had played out to its inglorious end. As the white man so memorably rationalized it to the Native American of the Great Plains as he appropriated the once-common land to his exclusive use: 'It's your **Manifest Destiny**[24] to be wiped out by us; you are too close to Nature. You and your damn buffalo.'

The farmer's instinct to keep his land for himself was no longer inhibited by cultural ideals of common good simply because any such inhibition was no longer adaptive. Ownership was what was now culturally adaptive, and like so many adaptations it had ramifications.

The abandonment of hunting and gathering and its replacement by farming led directly to another profound change to society: that of leadership. It was a radical and inevitable change that happened quickly, i.e., within a few thousand years. It's also extremely resistant to modification and will, in my opinion, play an influential role in the structure of society for a long time to come. To understand why and how it came about we need to say a bit about

---

[24] Manifest Destiny was the nineteenth century American rationalisation that the invading Europeans were ordained by god to occupy the land and subject the native savages to their will.

what it replaced. Assuming that the societies of surviving hunter-gatherers echo reasonably well those of the recent past in their fundamental practises we can take their method of governance to reflect hunter-gatherer leadership in general. The outstanding feature is the absence of individual leaders, and of authoritarianism. What was selected for or culturally promoted was a flexible system that relied on collective rather than individual leadership. Decisions about when to move, where to go and the management of group activities were made by consensual discussion in which everyone had a right to express their opinion. Deference was paid to individuals who excelled at something; cultural knowledge, hunting, gathering, fighting, settling disputes, magic, perceived wisdom, and so on, and their views tended to be the ones ultimately adopted. Individuals who tried to assert authority over the group were denigrated. In fact, any individual who showed conceit or excessive self-esteem was disparaged; modesty was expected even of the most admired individuals.

This system of governance has attracted a lot of attention and much rather adoring praise of hunter-gatherer egalitarianism. However admirable we may judge it, it was nevertheless just another adaptation selected for the advantages it conferred – advantages (whatever they were) that disappeared along with virtually the whole hunter-gatherer style of life. What replaced it was, in my view, a regression to a much more fundamental, instinct-driven form of governance; namely individual dominance. This is the primitive form of leadership as displayed today by our fellow apes, with the partial exception of bonobos in whom the copulatory instinct has been co-opted to mitigate violent physical dominance.

The hunter-gatherer group was close-knit, small and mobile, and selection for consensus leadership had clearly been strong. Farmers could not move to avoid competitors, and being widely scattered could not gang together for defense. Nor had they the time. It seems to me inevitable that to succeed as a farmer you had to dominate those who depended on your surplus, not only your kin but as many of your kith and kind as possible. There was a mutual need for an aggressive– and at the same time capable - leader who could mobilize the required labor from the non-farming people, not only for farming itself but far more importantly to secure the whole domain of the group of kith and kin. The genes for the instinct to dominate those around you had been fixed in us millions of years ago; it did not require much adaptive advantage to revitalize the effects of those genes. As surplus inevitably led to, *inter alia,* more surviving offspring competition for farmable land rapidly increased resulting in ever-increasing numbers of landless people. The more of these that the dominant individuals could mobilize the more secure their position. Thus was the landed gentry born as was the urban life that, ultimately, depended on their wealth. The birth of the landed gentry was also the birth of royalty, for the two require each other. As individuals amassed more and more control over surplus and expanded their spheres of control they generated the need for a second tier of dominance – one that controlled whole fiefdoms. The barons needed kings. For the first time in our history (as far as archaeological evidence goes) we had kings. And kings of course sought to dominate other kingdoms and ultimately, as is the situation today, the whole world. The rattling of sabers has never been louder. There is a forlorn, simple destiny underlying this process; being just apes our most powerful instincts are relatively simple ones. Much of civilization is, ironically, a regression to raw instinct.

One other consequence of dominance-based leadership was in my view the birth of slavery. I think it began, i.e., became adaptive in evolutionary terms, quite simply as a result of competition among leaders. As soon as you successfully invaded another baron's territory you became willy nilly responsible for a lot of alien people who survived the fighting. A common solution (now as then) was to starve or slaughter them. But, another solution was to take them captive and force them to work for you. Thus was slavery born, and prohibition notwithstanding, survives to the present day. I cannot see that anything more complicated than that is needed to explain slavery. One has only to look at the extraordinary contempt in which contemporary dictators hold their own people to realize that captives would only be viewed in the context of their potential utility.

10.2   Unification, Coercion and Cooperation

It suits most people to cooperate, as long as those whom it doesn't are forced to

The Myth of Cooperation

One of the most vaunted characteristics of us humans is our alleged cooperative behavior. Our admiration for this feature is in exact proportion to our fear of raw instinct. It reveals our awareness of the primary motivation of all individuals, which is to better themselves at the expense of others. Zoologists call it competition; others selfishness. From the outset of our evolution we have been social animals living in groups of closely and distantly related individuals. Living in groups cannot of itself relieve the individual of his responsibility to feed himself and ensure his safety so that he must still employ his basic instincts and compete with his compatriots. This basic tension between the individual's assertion of his needs and the disruption this can cause to the integrity of the group has been there from the beginning and has led to the evolution of mechanisms to dispel it.

Group integrity was first maintained by simple physical violence. While this is still widely practised by us as well as by other animals it has obvious drawbacks. It's costly, results in casualties and depends on fear, which must be constantly reinforced. In this respect, there's scant difference between a troop of chimps and a throng of hunter-gatherer humans – or pastoral savanna nomads of a few generations ago. Like the strong force of physics the individuals of both species are strongly held together by the bond of kind, of species, the nuclear component of the group structure. A weaker, but still powerful, instinctive force binds kith and kin. The third force is that which governs the day-to-day behavior of the group's individuals, in the case of chimps still instinctive, in us a largely conscious system of written and verbally perpetuated rules. One can imagine an analogue of the fourth and weakest force of Nature – the gravitational field – as a reaction by the group as a whole to all other groups in the environment. The difference is primarily

> The one means that wins the easiest victory over reason: terror and force.
> *Hitler*

that in chimps the third level of bonding remains as it was in us a million years ago – purely instinctive. Everything a chimp wants from its fellows is expressed instinctively, and, according to circumstance, modified or curbed by instinctive, inhibitory brain functions. We have largely replaced the instinctive regulation of instinctive wants by conscious rules, transmitted and enforced verbally and only where necessary, physically. Both systems work well – we and chimps have been zoologically successful for millions of years. But, chimps will not survive much longer, because, as hunter-gatherers, they can only thrive in the *terra nullius* we covet and will inevitably seize. Their Manifest Destiny, and that of all other large animals is exactly the same as the Native Americans' or the Hadza of Tanzania or the Kalahari Bushmen or the orangutans of Indonesia.

The way cousin chimp gets what he wants from another chimp who doesn't want to cooperate is by cunning or violence. We still use those methods, but our expanded consciousness enables us to modify them so as to reduce their deleterious side effects. We call this cooperation, misleadingly so, because cooperation is the perceived outcome, not the method. The notion of cooperation is a classic rationalization, an idealization of something much more fundamental: coercion. Having coerced a number of individuals into submitting to a common task (for the benefit of the coercer) we perceive an ancillary outcome – a benefit for everyone – and suppose it was the reason for the initial coercion. We call it cooperation. But, it's just a perception, because the term cooperation is benign, whereas coercion is aggressive, i.e., the term cooperation is a euphemism for coercion. Both the act of coercion and the softening of it by euphemism have been selected for good reasons: both confer advantages albeit of quite different kinds.

Historians, archaeologists, anthropologists all talk of the emergence of human cooperation as something fundamentally novel, something that replaced what went on before. They see the many instances of humans acting collectively to do things sometimes on a large scale that seem to subsume the individual's self-interest. We form queues, refrain from hitting each other, keep off the grass, dress uniformly. We pay taxes, join armies when told to, respect each other's property and beliefs, and so on. Knowing – unconsciously perhaps – that all these things are contrary to what the individual really wants people are impressed by what seems to them the high degree of altruistic cooperation. The perception is an illusion, a classic example of rationalization and idealization. What they are really looking at is the same old process of coercion backed by enforcement different only in that overt physical violence has been replaced by more subtle methods of enforcement. I'm speaking of course of western democratic societies; overt physical violence is still the primary method of coercion in dictatorships.

Western democracies that have eschewed overt physical violence have replaced it by the rule of law – the base on which the whole democratic ideal rests. But, the rule of law is still coercion. It prohibits the individual from using physical violence or inflicting punishment by transferring the right to enforce and punish to a putatively independent and objective justice system. Everything that rulers want done is couched in laws that oblige the people to obey. Physical punishment of those who defy cooperation is meted out by a putatively objective judiciary that is supposed to reflect society's ideal of fairness rather than the dictator's whim. Force is still the means by which obedience i.e., cooperation is brought about. Break the law and you will be subject to the ***full force*** of it, as we are constantly threatened on TV by our policemen and rulers. Without the rule of law and its continuous enforcement in every sphere of 'cooperative' life, society would rapidly disintegrate. We can't stage a rock concert or a football match or a court case or a teenage party without police to enforce cooperation. It just doesn't come naturally. Might remains right.

There is one more device that came to be used in the large-scale coercion of people: religion. In fact, it's the best device we've come up with, even though coercion was not the root of religion. There is nothing, though, that more efficiently binds large numbers of people together. Until recently, for all practical purposes, it bound whole nations together across the entire 'civilized' globe. A nation's identity was synonymous with its religion. Other things, notably language and geographical origin, were extremely influential, but none had the same valency as religion. Nowadays, political ideologies – notably socialism - have in some cases supplanted religion. What is conspicuous in such cases is that they have to be forcibly imposed and sustained. Religion's strength was that its members eagerly embraced it, and still do. Even though in the contemporary West around half the population no longer claims to be religious they nevertheless behave as if they were still Christians. They remain afraid of god and the church and avoid blasphemy or criticism, worried that it might disqualify them from the niggling, wishful possibility of immortality that still hovers in the backs of their minds.

The question arises as to why this is so. The strength of an ideal paradoxically lies in its unattainability. All of society's most longed for ideals – democracy, freedom, justice, peace, etc. – are essentially out of reach, quite possibly forever. Religion meets this criterion more than any because immortality will not take place in this world. It resides in the supernatural that by definition is beyond our ken. Only belief in god will get you a passport to eternity.

****

The notion of cooperation is one of civilization's most revered shibboleths, something that is taken for granted, a defining attribute. It's one of the principal weaknesses of the historian's method of explaining the ontogeny of civilization that they proceed by listing the various shibboleths of their trade; the time-honored narrative from regular ape to bipedal 'hominin', the 'revolutions' of religion, stone tools, domestication of plants and animals, cooperation, settlement, metal-working, literacy, science and so on. This gives a picture of history, and if the historian is a good writer it can be a beguiling one. It's always superficial, though, a procession of venerable rationalizations that produces a sort of periodic table of the elements of human history. It does not tell us *why* we hopped along these stepping stones, what went through our minds that led us to do what we did. The historian can never tell us our history until he can tell it biologically, the natural history of an animal. For we are simply one of the many animals that has evolved on this planet. There is nothing whatsoever about us that is not the product of evolution by the natural selection of genes that enabled us to exploit more of our surroundings and be more fecund. The self-marveling variety of our newly-evolved intellectual adaptations is as devoted to pillaging Nature and heedlessly multiplying as ever plain instinct was.

10.3   Reading, writing and arithmetic

It's necessary to put literacy in the context of our overall evolution because we are apt to take it for granted as being part of what makes us different to other apes. But, most of what makes us different evolved long before literacy. It's in fact extraordinarily recent. Only 150 generations, 5,500 years, ago, Sumerians in southern Mesopotamia invented writing to keep accounts of taxes, property and debts, i.e., a tool to aid in the tracking of surplus and wealth. Such accounts were kept on clay tablets inscribed with numbers – initially there was no other information. Within a thousand years or so the Sumerians had developed a comprehensive script – Cuneiform – but still written on clay or stone. At around the same time, the ancient Egyptians developed a comprehensive system of hieroglyphs.

For numbers Sumerians devised two systems: one to base 10; and the other to base 60. Base 10 presumably reflected the 10 fingers that must have been, as they are yet, used as counters. The choice of base 60 shows that considerable thought had gone into such matters, as 60 is a highly composite number with advantages in calculating fractions. A method of counting on one hand, still used, is to touch each finger sequentially with the thumb of one hand and count in dozens. We still use Base 60 for time and degrees of arc.

What is particularly noteworthy about literacy is that the earliest writings are as conceptually sophisticated as anything that came later. This means that although we only have a written record about 4,500 years old people had obviously thrashed out intellectual matters long before that; our intellectual capacity was fully formed ages before. The point I wish to emphasize is that literacy played no part in the evolution of our intellectual capabilities. Once paper had been invented it enabled a huge expansion of the quantity of what we thought about that made it easily accessible to other people, and as such was an important contribution to civilization.

There's an awful lot of ballyhoo about language. The received opinion that it's what distinguishes us from our fellow animals seems to me absurd. All animals have language, even bacteria. Language is the exchange of information. The medium can be chemical or physical - commonly the radiation of odors and sound. The only way our language differs from other animals' is that it evolved into an elaborate brainwashing tool to pass on customs or other information that wasn't instinctual. That bit of evolution was part and parcel of the expansion of our brain that enabled us to behave in ways, not yet fully understood, that gave us advantages over all other competitors as well as all other animals bigger than a rat. It's our brain expansion and concomitant widening of abstract consciousness that distinguishes us from other animals. The sheer capacity of our language is dazzling, and like many shiny things, misleadingly so. It's the information it can transmit that matters, whether it be utilitarian, artistic or supernatural. Instincts take millennia to evolve. Language was a tool that helped overcome instincts such as aggressive violence and redirect them on to some more remote social goal. Selection for spoken commands as substitutes for violence seems to me to have been the likely selection pressure on the initial expansion of our linguistic faculty, though various other pressures have been posited. Whatever the origin, it opened the way for our elaborate management of instinct by imposing cultural dictates verbally. This allowed for a rapid adjustment of customs to changes in the environment – a huge advantage where instinct alone might evolve too slowly to adapt to sudden, or profound environmental changes. Language became a powerful cultural device to augment group unity – a unique language is one of the characteristics that define a group sharing a distinct genome and culture.

Ultimately the utility and flexibility of language was extended to the exchange of ideas. Language doesn't precede ideas, or produce ideas – they come first. Cousin chimp can have quite complicated ideas without more than a rudimentary language, as can elephants and dolphins and no doubt other animals. In the case of our greatly expanded languages the evolution of an idea once it was first expressed would have been greatly facilitated by discussion. None of the things we do are the ***product*** of language, however much they may depend on it for their realization. Without language it would be difficult if not impossible, for example, to convey the relativity of matter, motion, time and space; but language of itself is no part of the concept.

****

That brings me to the end of the evolution of our instincts and their cultural modification. I've tried to emphasize that the fixing in our genomes of our suite of instincts encompasses a staggering length of time – millions upon millions of years. In comparison, the cultural management of these instincts began only yesterday in evolutionary terms. Whereas the cultural manipulation is adapted to relatively quick responses that are not incorporated into our genomes, but temporarily superimposed on them, the instincts themselves are left intact. Instincts are forever; culture changes with the times.

I finish now with a summary of what I've described and some remarks upon the consequences of being made of timeless instincts muffled by a façade of plastic culture.

Chapter 11   After all, we are only animals

The mother of all our problems: the pathetic fallacy that we humans are no longer animals.

Drinking when we are not thirsty and making love at all seasons, madam: that is all there is to distinguish ourselves from the other animals.
De Beaumarchais, Le Marriage de Figaro

I woke up on the roadside, daydreaming about the way things sometimes are
Visions of your chestnut mare shoot through my head and are making me see stars
You hurt the ones that I love best and cover up the truth with lies
One day you'll be in the ditch, flies buzzing around your eyes
Blood on your saddle
Dylan

I began this essay by saying how difficult I had found it to grasp why reason, a highly developed attribute of our minds, played little or no part in the really important things we chose to do. It took me a long time to see where I'd gone wrong, but when I did it all seemed so simple and obvious. My problem was that I had stubbornly clung to my cherished illusion that reason must ultimately prevail. I had fallen for a belief, a belief in the power of reason. Ironically, reason itself should have told me that instinct is still in charge, that for all our cleverness we remain animals. Animals responding like all others to their ancient, irresistible instincts and ultimately, regardless of all the tricks we use to fool those instincts, obeying them.

Naturally, as a zoologist I was embarrassed by this revelation: a zoologist should have known better all along. What we take to be the civilized aspects of our behavior all too easily obscure the basics, even for those who are well aware of it. A little thought showed me that it was easy to see why I had been so puzzled; why others – the majority in fact – seemed to place so little value on reason when it really matters. All I had to do was single out the main features of our evolution and describe the outcome – us humans – in plain zoological, i.e., evolutionary terms. The result made it clear that something as recently evolved as reason – at least abstract reason – would be most unlikely to displace millions-of-years-old ego-driven instincts – the very instincts that kept us alive all that time. Because those instincts are as healthy and vigorous as they've ever been. Sure, our 'higher order' mental faculties can do a lot for us, but we do not let them take over fundamental decisions such as how to maximize and above all control surplus (aka wealth). Circumstances change, but the principles underlying the maximization and control of wealth stay the same. And those principles really all boil down to plain old aggressive competition, the same principle behind all evolution.

Before making some concluding remarks I will summarize the steps I took in this analysis. I began by describing the strongest elements of our behavior according to a Freudian-based model of our minds in which the interplay between our unconscious and conscious minds is the primary mechanism that determines how we behave as individuals. I substantiated this by comparing what we know of brain anatomy and function to the precepts of the model and showing that there's no conflict between the model and what science has demonstrated so far by experiment. This is important because the model is based on observation, not experiment, and ultimately observation must be tested against experiment. More remains to be explained than has already been explained; but we already know enough to understand the basics. I emphasized the mantra of evolution, namely that all organisms compete with one another primarily for food and reproductive opportunities. Those that do best in this competition are those whose genes best fit them into the environment of the day.

I drew attention to the antithetical nature of our mechanism for understanding the world. I emphasized the primacy of the ego, our sense of self, that everything we do is for our individual pleasure where benefit and pleasure are synonymous. This fundamental fact of nature – the Pleasure Principle – has a sound zoological basis that becomes obvious when the genetic basis of evolution is taken into account.

I then described the most important defense mechanisms that evolved to protect the ego from reality when reality threatened to weaken or dismantle the ego's efforts to prevail. These defenses essentially enable the ego to bluff its way through life. They include lying, denial, projection, rationalization, and regression. An indirect defense is sublimation, whereby strong-willed individuals can re-direct their instinctive urges into creative actions – notably art, but also invention and in fact any imaginative or constructive activity.

Next I highlighted the instinct to idealize our strongest wishes, to create ideal objectives that are essentially unattainable because they transcend reality. Virtually all of society's most wished-for states – peace, tolerance, democracy, happiness - are ideals; clearly perceived but never fully attained.

I then drew attention to our two most powerful and influential instincts – aggression and copulation -that reflect the two fundamental strivings of all animals: for food and reproduction. Aggression is the manifestation of the often unconscious primal urge to compete with each other; copulation for the requirement to reproduce. Constant copulation evolved to ensure that undetectable ovulations continued to be fertilized. Making copulation so pleasurable that it became an end in itself ensured that vital constancy.

At this point (Chapter 2) I returned to the description of human behavior with an account of how we behave in groups within society as a whole. Such groups form around a social principle, new ideology, or popular theme. Individuals in such groups characteristically abandon most of the social inhibitions demanded by society and behave in a more naïve, simplistic and uninhibited way. They become susceptible to suggestion and slavishly obey group leaders' demands.

I then described (Chapter 3) brain function, specially as it relates to the stimulation and inhibition of fundamental behavior and the formation of memories. I ended Part 1 with a description (Chapter 4) of genes and natural selection.

The next step I took (Part 2) was to describe our evolution, beginning with our early evolution. Since I regard it as impossible to understand our behavior without knowing how it all evolved I have spent the greater part of this book tracing our collective adventures in Nature over the 19 million years it took to make our species. It's also a fascinating story in itself quite apart from its importance in understanding our current society and its directions. It's easy to forget that the theory of evolution by natural selection – the basis of zoology and one of the most robust theories we've ever come up with – is brand new, a mere 160 years old. That we had no idea of what made us what we are for almost our entire existence led us into a vast amount of misconception, loony ideas and pure baloney and is one of the reasons why our current activities are so bizarre. Not only that, but in the short time since we discovered why we are as we are, a huge chunk of our population never gets to learn how it works; either because school curriculae don't include it – even reject it - or because schooling for them is rudimentary or non-existent. Even the average economist or lawyer or politician has probably never been formally educated about our evolution. While I find that amazing it's just one of the many amazing choices we make.

(Chapter 5) describes the origin of bipedalism and the appearance of our kind of ape, the australopiths.

Part 3 covers most of the Pleistocene (Chapter 6), the famous Stone or Ice Age during which we evolved most of the features that define us. Part 4 (Chapters 7 and 8) covers the last 600,000 years of the Pleistocene that saw one of the most outstanding features of our evolution – the Big Beautiful Brain. During this phase our brain grew some 17% bigger, while the rest of our anatomy scarcely changed. To accommodate the bigger brain our craniums grew larger and more globular.

In Part 5 I discussed (Chapter 8) the evolutionary origin of one of our most influential characteristics; our capacity for belief, particularly belief in a supernatural underworld and how this developed into belief in an afterlife as a result of the fear of death. I went on to show how religion evolved from the supernatural, the ritual worship of and subjugation to specific gods, culminating (in the West) in a single fatherly god. I drew attention to a curious peculiarity of our species – its sense of rhythm and the music that grew out of it. The selective value of rhythmic music is as a coercive device that unifies people in a common, highly pleasurable activity.

I then described what I consider the single most important mental outcome of the Big Beautiful Brain phase: our conscience and its enforcer – our sense of guilt.

Part 6 (Chapter 9) covers the last 23,000 years of the Pleistocene, after brain size stabilized. We engaged increasingly in an activity that was to determine, more than any other, how our society would evolve into what we see today. This was the substitution of grass seeds for all other foods, and on learning how to grow these seeds. Within 23,000 years reliance on grain grown under our control had completely replaced hunting and gathering in the Mesopotamian Fertile Crescent, from where it soon (soon being a few millennia) spread to the rest of the world. In an evolutionary blink of an eye millions of years of hunting and gathering were tossed aside in favor of growing and eating grass seeds.

The fossil record suggests that Neanderthals had been eating opportunistically gathered grain for at least 100,000 years. Presumably we did too, we just don't know. Whatever the history of eating grain it wasn't until after the Big Braininess that it dominated our food supply. It did so for three main selective reasons: the far, far higher calorific return per unit of energy expended; the relatively easy accumulation and storage of surplus; and the paving of the way to permanent settlement. It has to be assumed that the prior brain changes included adaptations for the abstract projection into the future that cultivating and guarding certain grasses that wouldn't be edible for many long months entailed.

Part 7 deals with the Holocene. The shift to farming and becoming sedentary resulted in another fundamental change to society: the regression to a more primitive expression of ownership. Hunting and gathering had culturally suppressed the natural instinct to possession of the food you found in favor of communal access to it. Farming did the opposite, because ownership of the land on which the surplus grew was vital if you were to enjoy the enormous pleasures to be had from surplus; you couldn't do that if you shared it with everyone. Thus was the landed gentry born. Thus was the crime scene set for the rape of Old Mother Nature.

The first known villages of simple, oval structures go back 14,500 Y. Just over 5,000 years later we were making tightly-packed, strong, double-story, rectangular houses with dedicated storage rooms – an index of the extent of organized grain farming (animal farming was just beginning). A stronger index is that around this time, 9,000Ya, some of these villages – Jericho is the classic example – were walled. In other words, the raiding of villages by rival landowners – the bigger ones called themselves kings – was well under way. The basic structure of society right up to the present time was established by these rival kingdoms vying for power over people and wealth.

Another feature of this period was that kings began diverting a part of surplus into the building of huge structures unrelated to everyday living. The practical difficulties of quarrying, dressing and erecting the huge blocks and pillars involved must have been immense; the urge to do so must have been correspondingly strong. The expense must have been enormous. The purpose has traditionally been assumed to be for rituals to appease the gods of the day, because later on such structures were for exactly that.

The perceived advantages of farming over hunting and gathering led to the former's adoption all over the world and culminated in the relentless destruction of Nature everywhere to make way for it. The Rape had degenerated into the Murder that is currently in its final phase; at the present rate of destruction the last of the world's rain forest and most remaining species of large animals and plants will be gone in 80 years.

Farming and settlement forced one other influential change on the structure of society by creating large assemblages of people not directly related to one another. Those who were driven to control surpluses inevitably tried to increase their wealth by uniting smaller, hitherto unrelated groups into bigger ones. These leaders were called kings – dictatorial leaders of one sort or another forever contesting each other for the most dominating position. That basic structure of society of course is what we still have today.

The mastery of large assemblages of people inevitably trying to act in their own or a small group of kin's interests was initially by force – as it is in cousin chimp's society and still is in a majority of today's societies. Gradually, though, direct physical force was replaced by the rule of law, which transferred the control of force from the leader to a putatively independent and impartial justice system. This didn't hinder dictators – it made it easier for them, because they made and administered the law. In Western democracies the law is putatively independent of politics, operating without fear or favor of anyone. It's an NSDT statement that that ideal remains a long, long way off, even though there is nothing but political interference in the way.

I then discussed the myth of cooperation, how we confuse coercion with its outcome – cooperation. The control of society was originally by violence, as in cousin chimp and our dictatorships today. Its replacement, the rule of law, still operates by the threat of violence in the form of imprisonment. Fear of punishment still is the way we maintain so-called cooperation. We have to police every group event we stage from rock concerts to football matches to elections.

I emphasized the huge role religion played as a powerful coercive device in that it could unite any number of unrelated or related people under a single ideal. Furthermore, that ideal was more powerful than any other ideal because it could not possibly be realized in this world; it belonged to the supernatural world that can only be entered by believers after death.

I discussed the important part played by language, and recently, literacy in the application of cultural controls of instinct.

Finally, in this last chapter I try to make the point that we are only animals, not some sort of superior being or divine confection, and that as such we behave more in accordance with our ancient, animal instincts than with the cultural restraints and ideals imposed on those instincts.

****

I've said that the mother of all our problems is the pathetic fallacy that we humans are no longer animals. We dismiss our cousins - chimps, bonobos, gorillas and orangutans -as inferior apes because we think we are controlling our instincts – something they can't do. It's true that we have evolved some ability to take the edge off our instincts, but that is a far cry from control or mastery. We still have and are driven by the same old instincts; we remain the same five apes we always were. If not, exactly where and when did we transmogrify into something not animal? Surely there must have been an outstanding, easily recognizable event that suddenly made the switch that shed millions of years of beastliness to leave like an emerging butterfly the grubby chrysalis of our former life behind. All I can see are millions and millions of years spent as apes gradually evolving into slightly different apes all doing as far as we can tell much the same thing. The only event that radically changed what we do was the switch from foraging to farmed surpluses that by 10,000 Ya was widespread in Asia, and by 7000 Ya in Africa, and by 4,000 Ya had got to Europe too. Surpluses freed us from having to spend all our energy on finding today's food. We now had the time and tradable wealth to coerce ever larger numbers of people into producing ever more surpluses of increasing variety to create ever more wealth – as we currently are obsessed with doing. Did that mean we were no longer animals, but rather something that still *looked* and behaved exactly like yesterday's ape, but which was nevertheless not an ape any longer, but a being superior to and better than apes and all other animals? You cannot work that out scientifically or logically or any way based on demonstrable facts. You have to fall back on the age-old magic remedy of belief. Oh dear, oh dear, oh dearie me.

We none of us take kindly to insults, to being disparaged. It's an old but still popular insult to call someone a monkey and few would not bristle at being so called. Yet somehow I have given myself the task of justifying why I call us all apes (at least a step up from monkey). There is the popular myth, inadvertently ignited by Darwin himself, that we are descended from apes. What descent? There is no point in our evolution that we can single out and say 'Look, there it is, the exact point at which we left apedom and became beings.' What do we expect from that mythical point? What sort of attribute is it that would define the transition? In what way is a being distinguishable from an animal? What is it we descended from that the other apes are still bogged down in? For a zoologist there's no such singular point, nothing that would unequivocally distinguish us from our fellow apes. There are things we do that cousin chimp doesn't (like stuff up our habitat); insofar as they are the products of natural selection they are of the same texture as all other behavior.

I think it's safe to say that this mythical point is signaled to those who believe in it by the acquisition of our so-called 'higher' attributes. We have evolved the capacity to tone down our selfish instincts and sublimate their energy into 'unselfish' behavior that enhances the productive capacity of the group we call society. It does so because it facilitates peaceful and harmonious behavior among many where before the immediate interests of the individual took precedence. It's effectively inclusive fitness on a large scale and encompasses what we know as civilization. The problem arises because most people see this as something unique to us, something that elevates us from the beast. For a zoologist, though, it's simply another product of natural selection, of favorable attributes that arose by chance mutations. And whether or not it's unique to us depends on your view of the extent to which other highly evolved animals may have acquired elements of the same behavior.

The point I wish to emphasize in all this is that regardless of the importance of our 'higher attributes', and of our idealization of them, they do not **replace** our primary instincts; they only soften them. They set up the façade. All our behavior is still initiated by our instincts. There is an NSDT element to this; namely, that higher order behavior does not arise spontaneously or independently; it is by definition an alteration of a prior, instinctive urge.

> I wish it, I command it.
> Let my will take the
> place of reason.
> *Juvenal*

There is an obvious obstacle to defining us as animals driven primarily by instinct and that is that today's urban society has for generations grown up knowing only a man-made environment that contains nothing but fragments of the natural environment in which their ancestors evolved. The same even goes for the greater part of contemporary rural society. Such people have to imagine Nature – they've never seen it, let alone felt part of it. The problem is that while their environment has changed radically they are unaware of that fact, and naturally define themselves in relation to the environment of the moment. And their definition is inevitably lacking in what might be called the animal elements focusing instead on our 'civilized' features. Maybe it doesn't matter, that what the irrelevant majority believe is of no great moment because they don't participate in what their leaders decide is best for them. After all, it's how the leaders make their decisions that counts.

What I'm leading up to, what I'm trying to make self-evident, is not so much that we remain animals as that there are consequences to being animals that explain quite simply why we are stuck with the great why whys that are the subject of this essay. All our leaders, whether dictators or 'democratically elected,' unconsciously commit themselves (and therefore us) to increase – increased production driven by increased producers. Ever-expanding economies powered by ever-expanding populations. More, more, more. The outstanding feature of this is that it is never questioned or explained or justified. Much time and trouble (even reason) is spent drawing up annual budgets for the realization of economic expansion, a modicum of reason is used to figure out policies and procedures; but **why** we are doing it is never discussed, never explained, never related to a plan for our future. Much is made of our capacity to think ahead, to plan, to arrive at a clearly defined goal. Yet these faculties play no part whatsoever in our frenzy to expand.

There can be only one explanation for this extraordinary behavior - that it's **purely** instinctive. The work of naked, raw instinct. We mask our primitive instincts with a thick mantle of inhibition and redirection, designed – quite consciously a lot of the time – to make them compatible with our ideal of an orderly, peaceful and civilized society. Two instincts in particular have long been subjected to this domestication of their original wild, spontaneous expression; the aggressive and copulatory instincts. The drive to copulate is disciplined to function privately and not disrupt public order. The aggressive instinct is denied to individuals for violent, personal use and channeled instead into socially acceptable forms of competition (violence is reserved for state use only). And that's the key to it - **socially acceptable forms of competition.** And there's nothing more socially acceptable than generating more surplus, more wealth; nothing, absolutely nothing.

Leaders who concentrate on aggressive competition that is overtly designed to increase the prosperity of the led will always be followed. They do not have to justify their behavior – it's self-evidently the best thing they could be doing. Like religion, it's a belief so old and deep-seated that even if you have doubts you dare not voice them. And, if we are to be objective about all this, there is the most iron-clad of all instincts supervising everything, but **everything** we do: the ego, as we awkwardly call it. It's an NSDT statement that the ego comes first, that its satisfaction, i.e., its competitive success, sets the rules for all individuals' actions. We never promote actions that are not primarily driven by self-interest, however much trouble is culturally taken to hide that fact by disparaging it as selfish, self-serving, narcissistic or simply egoistic. Culturally, we are expected to ignore the basic drive and attribute some more selfless or altruistic motive. That's okay, of course, because civilizing raw instinct is what we spent a considerable part of our evolution doing. My point is that we should never forget what's really driving us, because the individual naturally finds it hard to curb his ego, to satisfy it with more distant or less personal rewards. The most conspicuous expression of ego is power. Leaders are driven by what to them is the sublime taste of power over others that we the led can only marvel at. They can't help it; the genes are dictatorial. They create dictators, and all leaders would be dictators in their ideal worlds. All that we the led, the irrelevant majority can do is try to make our leaders use their addiction for our benefit as well as theirs, however improbable that ideal may be.

The unquestioning belief in ever-expanding economies is an expression of our age-old aggressive competitiveness that does not call for any inhibition. It's simply plain raw instinct doing what it evolved to do. To a zoologist a clear signal repeats in the otherwise noisy data that describe today's civilization: the amassing and control of surplus. The behavior that natural selection fixed in our genome back when we were beginning to grow grass-seeds and accumulate surplus food. It would be regarded by most – leaders and led alike - as astonishing were this belief to be scrutinized or questioned. Somehow it lies outside the terms of reference of our higher order faculties, because, I suppose, it *seems* to raise no social problems. It's easily rationalized and fulfills a universal ideal – the ideal of unlimited wealth that itself claims to guarantee unlimited happiness. There is a strange sort of irony in the unconscious rejection by leaders of reason or any sort of enquiry when it comes to economic and population expansion. It's as if they are saying 'We can't have people questioning the really fundamental things – they're much too important.' Which is the same thing as saying that an instinct as strong and basic as striving after more is far too important to be subject to such intellectual frippery as reason.

> Reason: Fools gold for
> the thoughtful.
> *Yann Martel*

The matter of the ever-expanding population is complicated by an element of taboo - the notion that limiting the number of offspring is off-limits. Contemporary society is preoccupied with the idea that nothing is more sacred than individual life, and further, that the individual must be guaranteed certain social entitlements. These include the familiar so-called democratic freedoms- of expression, from persecution, from slavery, etc. One of these freedoms is to reproduce spontaneously without interference from the state or anyone else. This freedom gains its strength from the simple fact that it suits leaders to have an increasing number of people because wealth is proportional to the number of people generating it, not just those of a given state itself, but of all states with which it trades. It's bolstered by monotheistic religion that celebrates the sanctity of the family – go forth and multiply. The Catholic version of Christianity for example forbids even contraception. It should be noted here that the fact that a large proportion of the world's people cannot even sustain themselves let alone produce surplus is only an apparent contradiction. They generate indirect production in the form of economic aid and military help.

Part of the unquestioning commitment to economic expansion is its momentum. It's what we've always done for the last 10,000 years or so, ever since we learnt about surplus. Everybody has always taken it for granted that the only way to keep your enemies - your competitors - at bay, to stay a step ahead of them, is to get bigger and wealthier. As long as everybody was on the same page this assumption was justified. Recorded history clearly documents the unending attempts by the leaders of the wealthiest, most populous states to overrun or dominate everyone else, a trend that continues unabated. The US, China, Russia and the EU vie for domination of the world. Less populous states vie for domination of others in their league (dreaming the while of when they get bigger and can afford atom bombs). It's a self-perpetuating situation that needs no input. Such things as reason were and are devoted to maximizing the components of the process, not the process itself.

I said earlier that the drive to increase, the lust for more, more, more seemed to raise no social problems. In its early millennia this was true because populations were still modest and there was all this poxy Nature crying out to be got rid of to make way for 'development.' Such

> The problems that exist in the world today cannot be solved by the thinking that created them.
> *Einstein*

people as might have thought about it saw only a lot of promise waiting to be taken up. It wasn't until just a few years ago - less than a century - that people realized monumental social problems *were* looming, hazily perceived through the miasma of accelerating prosperity. But, the sheer enormity of these problems was too much for people to grasp, or rather want to grasp, and instead they homed in on smaller, more manageable ones; pollution, atmospheric warming, civil rights, etc. These perceptions were all driven by the led, not the leaders (and there's nothing more tiresome for the led than trying to influence their leaders. It takes forever, because leaders are immune to reason and only give in when further resistance threatens obedience). If any leaders ever saw the looming monsters they fell back on the psychological defense of denial, creating the situation of today: what elephant in which room?

It's a strange thing that although we have been well aware of the vast difference between individual and group behavior for well over a century we rarely take it into account. The dross left behind by the idiot wind blowing through society includes much that is made by group behavior. Leadership, whether dictatorial or elected, conforms far more to group than individual behavior. The same goes for religious, corporate and protest or dissent group leadership. What we would call the civilized aspects of society are invariably the work of individuals. Society's groups are, by their genes, incapable of civilized behavior. Until we abandon, for example, party politics and devise rule by associations of individuals we are doomed to be governed by group behavior. Armies and religions have never of themselves added anything to civilization. Yet it seems that we are far, far away from being able to do without them.

As to where the galloping monsters of expansion are headed I'm reluctant to speculate. Are we really going to sort things out by fighting until only one mob survives and dominates the world? The main contenders are as bellicose as they've ever been and busy building up their armories. There's absolutely no doubt about what's on their leaders' minds. There is also the ever more apparent fact that democratically elected leaders are increasingly reluctant to deal with dictators in terms that can deter them. It's hard not to see their criticisms as mere appeasement. The truth is hidden behind an ever-thickening façade of euphemism, diplomatic double-talk and humbug. Statistically, dictatorships are on the increase. Nearly half the world's population is already ruled by outright or *de facto* dictators. What are the more civilized powers going to do about this if they won't fight (and in any case the led of today will refuse to be deployed as political cannon fodder)? I suspect nothing, except whine, wag the finger and rattle their sabers. It's said that he who hesitates is lost....

Are we really going to let the remaining forests be cleared (as we cleared ours, remember), to be replaced *in less than a century* with crops and towns? There's nothing whatever to suggest that those powerful enough to prevent it have the slightest desire to do so. Nor is it difficult to see why. Their leaders can't for the life of them see what all the fuss is about. Crops take up CO^2 and emit oxygen much as forests do and all those wild species of animals and plants contribute next to nothing to production. On the contrary, they impede 'development'. Leaders are obsessed with control, and pristine Nature is out of control. It's no good pointing out that clearing away wild animals and plants doesn't mean you've tamed Nature, because there remain the vast untamable aspects; cyclic climate change, cataclysmic crustal events, asteroid collisions, potent new viruses. For what political leaders can't see simply isn't there as far as they're concerned; it's a waste of breath to demand they consider the future rather than the present.

As for population increase, well, we leave that to its own devices. Always have, probably always will. Statistically, the number of children per family rises in proportion to their poverty, with the feedback effect of the number of offspring to be sustained exacerbating the poverty. Religion, to which the poor are particularly vulnerable, perpetuates the imbalance. The copulatory instinct cannot be quelled and managing it is only feasible for the wealthy. And all the while the wealthier, more powerful leaders bemoan declining birth rates and offer incentives to increase it.

What staggers me more than anything is that all this is done without any of the main players having the foggiest idea of what is driving them; instinct, plain ordinary old animal instinct. They delude themselves that it's the rational outcome of their thoughts and ideals.

****

Mother Nature spent half a million years supercharging our already highly evolved brain, but all we do with it is rape and murder her.

Only two things are infinite – the universe and human stupidity, and I'm not sure about the former.
Einstein

Acknowledgements

Thanks again to Jane for ideas, criticism and encouragement.

Photograph of the author by Jane Graham

Cover image and design by Jane Graham

Illustrations by the author

About the author

I was born and grew up in Kenya. I took an honours degree in Zoology from Rhodes University, South Africa, and later a master's degree from the University of Nairobi. My first job was with the then Kenya Game Department as a biologist; but, as funding for a biologist never showed up I worked instead as a game warden. After a great few years as a warden (mainly confirming John Lennon's observation that time you enjoyed wasting was not wasted) Ian Parker and I formed East Africa's first private wildlife consulting firm and spent the next ten years pioneering the aerial surveys that revealed for the first time the numbers of large wild animals that, in the sixties, occupied much of the country.

We also pioneered the active management of these animals, believing at the time that the only hope of conserving large wild animals was to make them at least as valuable as domestic livestock. It seemed that we'd hit on the way to perpetuate the lives of these animals forever. But, for many reasons it couldn't work in East Africa.

In 1973 I published *The Gardeners of Eden*, an essay on the psychological origins of conservation convictions, and with Peter Beard, the controversial *Eyelids of Morning* that attracted widespread attention.

I then worked on conservation projects in such Gardens of Eden as the Okavango, the Kalahari, Cross River National Park near the Cameroon border, Manovo in the Central African Republic, Omo, Mago and Nechisar in southern Ethiopia, Lake Bangweulu, the Bijagos Archipelago in Guinea Bissau, the Rufiji, the Ruaha and other spectacular places. Later I went to Papua New Guinea, Kakadu in the Northern Territory of Australia and Christmas Island.

I saw again and again that human population expansion is always going to be at the expense of Nature. The bushveld and its inhabitants are squeezed into ever diminishing patches of land. The Idiot Wind of civilization approaches gale force, and it's far from certain that we Gardeners of Eden will keep our flowers from withering away altogether.

Contact the author on LinkedIn

www.ingramcontent.com/pod-product-compliance
Lightning Source LLC
Chambersburg PA
CBHW061802250726
48657CB00001B/253